# Lecture Notes in Bioinformatics

Edited by S. Istrail, P. Pevzner, and M. Watern

Subseries of Lecture Notes in Computer Science

Marie-France Sagot
Maria Emilia M. T. Walter (Eds.)

# Advances in Bioinformatics and Computational Biology

Second Brazilian Symposium on Bioinformatics, BSB 2007
Angra dos Reis, Brazil, August 29-31, 2007
Proceedings

 Springer

Series Editors

Sorin Istrail, Brown University, Providence, RI, USA
Pavel Pevzner, University of California, San Diego, CA, USA
Michael Waterman, University of Southern California, Los Angeles, CA, USA

Volume Editors

Marie-France Sagot
INRIA Rhône-Alpes, UMR 5558 Biométrie et Biologie Évolutive
Université Claude Bernard, Lyon I
43, Bd du 11 novembre 1918, 69622 Villeurbanne cedex, France
E-mail: Marie-France.Sagot@inria.fr

Maria Emilia M. T. Walter
Universidade de Brasília, Instituto de Ciências Exatas
Departamento de Ciência da Computação (CIC)
Campus Universitário – Asa Norte, Brasília, DF, CEP: 70910-900, Brazil
E-mail: mariaemilia@unb.br

Library of Congress Control Number: 2007931833

CR Subject Classification (1998): H.2.8, F.2.1, I.2, G.2.2, J.2, J.3, E.1

LNCS Sublibrary: SL 8 – Bioinformatics

ISSN    0302-9743
ISBN    3-540-73730-8 Springer Berlin Heidelberg New York
ISBN    978-3-540-73730-8 Springer Berlin Heidelberg New York

Springer is a part of Springer Science+Business Media

springer.com

© Springer-Verlag Berlin Heidelberg 2007

Typesetting: Camera-ready by author, data conversion by Scientific Publishing Services, Chennai, India
Printed on acid-free paper     SPIN: 12093663     06/3180     5 4 3 2 1 0

# Preface

The Brazilian Symposium on Bioinformatics (BSB 2007) was held in Angra dos Reis (Rio de Janeiro), Brazil, August 29-31, 2007, at the Portogalo Suite Hotel. BSB 2007 was the second BSB symposium, although BSB is a new name for the Brazilian Workshop on Bioinformatics (WOB). This previous event had three consecutive editions in 2002 (Gramado, Rio Grande do Sul), 2003 (Macaé, Rio de Janeiro), and 2004 (Brasilia, Distrito Federal). The change from workshop to symposium reflects the increased quality and interest of the meeting. BSB 2007 was co-located with the International Workshop on Genomic Databases (IWGD 2007).

For BSB 2007, we had 60 submissions: 36 full papers and 24 extended abstracts, submitted to two tracks, computational biology/bioinformatics and applications. The second track was created in order to receive and discuss research work with a biological approach, and so to reinforce the participation of biologists in the event. These proceedings contain 13 full papers that were accepted, plus 6 extended abstracts. These papers and abstracts were carefully refereed and selected by an international Program Committee of 48 members, with the help of some additional reviewers, all listed on the following pages. We believe that this volume represents a fine contribution to current research in computational biology and bioinformatics, as well as in molecular biology.

The editors would like to thank: the authors, for submitting their work to the symposium, and the invited speakers Roded Sharan (Tel-Aviv University, Israel), Alberto Martín Rivera Dávila (Fundação Oswaldo Cruz, Brazil) and João Paulo Kitajima (Allelyx Applied Genomics, Brazil); the Program Committee members and the other reviewers for their support in the review process; the General Chair Sérgio Lifschitz and the local organizers Daniel Xavier de Sousa, Cristian Tristão and José Maria Monteiro; the symposium sponsors (see list in this volume); Nalvo Franco de Almeida Jr., João Carlos Setubal, José Carlos Mombach, Marcelo de Macedo Brígido, and again Sérgio Lifschitz, members of the Brazilian Computer Societys (SBC) special committee for computational biology; and Springer for agreeing to print this volume.

August 2007

Marie-France Sagot
Maria Emilia M. T. Walter

# Organization

BSB 2007 was organized by the department of Informatics - Pontifical Catholic University of Rio de Janeiro/Brazil.

## Executive Committee

Conference Chair
    Sérgio Lifschitz
    Pontifical Catholic University of Rio de Janeiro
    Brazil

Local Arrangements
    Daniel Xavier de Sousa
    Cristian Tristao
    Jose Maria Monteiro
    Pontifical Catholic University of Rio de Janeiro
    Brazil

## Scientific Program Committee

Program Chairs
    Marie-France Sagot
    INRIA, France
    *Computational Biology and Bioinformatics*

    Maria Emilia Machado Telles Walter
    University of Brasilia, Brazil
    *Applications*

## Program Committee

| | |
|---|---|
| Said S. Adi | (Federal University of Mato Grosso do Sul, Brazil) |
| Nalvo F. Almeida | (Federal University of Mato Grosso do Sul, Brazil) |
| Alberto Apostolico | (Accademia Nazionale dei Lincei and Georgia Tech) |
| Fernanda Baiao | (UNIRIO, Brazil) |
| Valmir C. Barbosa | (Federal University of Rio de Janeiro, Brazil) |
| Ana Lúcia Bazzan | (Federal University of Rio Grande do Sul, Brazil) |
| Marcelo M. Brígido | (University of Brasilia, Brazil) |
| Edson N. Cáceres | (Federal University of Mato Grosso do Sul, Brazil) |
| André P. L. F. de Carvalho | (University of São Paulo-São Carlos, Brazil) |
| Maria Cláudia Cavalcanti | (Military Institute of Engineering, Brazil) |
| Dominique Cellier | (University of Rouen, France) |

| | |
|---|---|
| Laurent Dardenne | (National Laboratory of Scientific Computation, Brazil) |
| Alberto M. R. Dávila | (Fiocruz, Brazil) |
| Zanoni Dias | (University of Campinas, Brazil) |
| Carlos E. Ferreira | (University of São Paulo, Brazil) |
| Ana T. Freitas | (Technical University of Lisbon, Portugal) |
| Richard Garrat | (University of São Paulo-São Carlos, Brazil) |
| Raffaele Giancarlo | (Universita degli Studi di Palermo, Italy) |
| Katia S. Guimarães | (NCBI, USA/ Federal University of Pernambuco, Brazil) |
| David Huson | (University of Tübingen, Germany) |
| João P. Kitajima | (Alellyx, Brazil) |
| Gunnar Klau | (Freie Universität Berlin, Germany) |
| Gad Landau | (University of Haifa, Israel) |
| Melissa Lemos | (Pontifical Catholic University of Rio de Janeiro, Brazil) |
| Natalia Martins | (Embrapa/Biological Resources and Biotechnology, Brazil) |
| Wellington Martins | (Catholic University of Goias, Brazil) |
| Marta Mattoso | (Federal University of Rio de Janeiro, Brazil) |
| Alba C. M. A. Melo | (University of Brasilia, Brazil) |
| Antonio B. Miranda | (Fiocruz, Brazil) |
| Satoru Miyano | (The University of Tokyo, Japan) |
| José C. Mombach | (Federal University of Santa Maria, Brazil) |
| Nadia Pisanti | (University of Pisa, Italy) |
| Alexandre Plastino | (Federal Fluminense University, Brazil) |
| Leila Ribeiro | (Federal University of Rio Grande do Sul, Brazil) |
| David Sankoff | (University of Ottawa, Canada) |
| Luiz F. Seibel | (Pontifical Catholic University of Rio de Janeiro, Brazil) |
| João C. Setubal | (Virginia Bioinformatics Institute, USA) |
| Roded Sharan | (Tel-Aviv University, Israel) |
| David Sherman | (CNRS, France) |
| Siang W. Song | (University of São Paulo, Brazil) |
| Marcílio C. P. de Souto | (Federal University of Rio Grande do Norte, Brazil) |
| Osmar Norberto de Souza | (Pontifical Catholic University of Rio Grande do Sul, Brazil) |
| Guilherme P. Telles | (University of São Paulo-São Carlos, Brazil) |
| Cristina Vieira | (INRIA, France) |
| Sérgio Verjovski-Almeida | (University of São Paulo, Brazil) |
| Martin Vingron | (Max Planck Institute, Germany) |
| Michael Waterman | (University of Southern California, USA) |
| Fernando von Zuben | (University of Campinas, Brazil) |

# Additional Reviewers

Christian Baudet
Markus Bauer
Luciano Digiampietri
Alan Mitchell Durham
Cristina G. Fernandes
Ivan Gesteira Costa Filho
Alexandre Paulo Francisco
Ronaldo Fumio Hashimoto
Dennis Kostka
Alair Pereira do Lago
Helena Cristina Gama Leitão
Ana Carolina Lorena
Simone de Lima Martins
Mariá Cristina Vasconcelos Nascimento
Christian Rausch
Christine Steinhoff
Yoshiko Wakabayashi

# Sponsoring Institutions

Brazilian Computer Society
CAPES

# Table of Contents

## Selected Articles

## Extended Abstracts

# Automating Molecular Docking with Explicit Receptor Flexibility Using Scientific Workflows

K.S. Machado, E.K. Schroeder, D.D. Ruiz, and O. Norberto de Souza

Laboratório de Bioinformática, Modelagem e Simulação de Biossistemas - LABIO
Programa de Pós-Graduação em Ciência da Computação, Faculdade de Informática, PUCRS,
Porto Alegre, RS, Brasil
{kmachado,eschroeder}@inf.pucrs.br,
{duncan,osmar.norberto}@pucrs.br

**Abstract.** Computer assisted drug design (CADD) is a process involving the execution of many computer programs, ensuring that the ligand binds optimally to its receptor. This process is usually executed using shell scripts which input parameters assignments and result analyses are complex and time consuming. Moreover, receptors and ligands are naturally flexible molecules. In order to explicitly model the receptor flexibility during molecular docking experiments, we propose to use different receptor conformations derived from a molecular dynamics simulation trajectory. This work presents an integrated scientific workflow solution aiming at automating molecular docking with explicit inclusion of receptor flexibility. Enhydra JAWE and Shark software tools were used to model and execute workflows, respectively. To test our approach we performed docking experiments with the *M. tuberculosis* enzyme InhA (receptor) and three ligands: NADH, IPCF and TCL. The results illustrate the effectiveness of both the proposed workflow and the implementation of the docking processes.

## 1 Introduction

One of the most important features of Bioinformatics is the collection, organization and interpretation of a large amount of information [1]. To carry that out, different computational tools have to be used to manage different data elements in a particular sequence of computational steps. Generally these kinds of experiments, called *in silico*, involve a sequential execution of a number of computer programs, where the output of one is the input for the next.

Usually these programs are executed in a one-by-one basis either manually or by simple shell scripts specially designed for this purpose. Often, one has also to face problems with the heterogeneous and distributed nature of the generated data due to particular input/output formats from the available tools [2]. Furthermore, manual execution of computational programs or the use of shell scripts to execute them usually lead to problems related to computer program diversity of usage, data flow recording and process maintenance. An interesting approach to model these characteristic problems is by using scientific workflows [3]. These workflows involve sequences of analytical steps that can deal with database access, data mining, data analysis, and many other possible steps involving computationally intensive jobs.

M.-F. Sagot and M.E.M.T. Walter (Eds.): BSB 2007, LNBI 4643, pp. 1–11, 2007.
© Springer-Verlag Berlin Heidelberg 2007

From Biology we know that macromolecules (receptors), such as proteins, enzymes, DNA, and RNA, are not rigid entities in their cellular environment. Therefore, it is highly desirable that this flexibility be explicitly considered during a computer assisted drug design (CADD) process. One important step of the CADD procedure is the molecular docking, where the binding of a small molecule (ligand) to its receptor is computationally tested and evaluated.

Docking experiments can be performed by a number of docking simulation software [4]. Most of them can deal with ligand flexibility, but have difficulties in handling the receptor flexibility. Those capable of handling receptor flexibility do it only in a limited way [5]. In order to overcome this problem and include a more realistic representation of the receptor flexibility during docking experiments, we considered an ensemble of thousands of receptor conformations, generated by molecular dynamics (MD) simulations. For each receptor's possible conformation, one ligand docking experiment has to be performed and analyzed. These steps are currently being executed manually, where the appropriate parameters (such as the protein and ligand names, the names of the files from the MD trajectory, the number of MD snapshots, the docking parameters, etc.) as well as the sequence of execution are defined by the user. Consequently, to re-execute a process with different receptor and/or ligand, the user would probably face serious difficulties to adjust the input parameters and data files. In addition, following and registering all execution processes are not simple tasks.

This article aims to model and automate the molecular docking process so as to explicitly include the receptor flexibility, and analyze their results. To accomplish that, we developed a scientific workflow, modeled using the JAWE design tool [6] executed by the Shark workflow engine [7]. The docking software AutoDock3.05 [4] and the PTRAJ module of the AMBER6.0 package [8], herein called PTRAJ only, are executed by scripts and computer programs written to perform each one of the activities described in the workflow.

This article is organized as follows: Section 2 presents a review of some basic concepts, important to understand this work, like the CADD process, molecular docking and MD simulation; Section 3 presents the developed scientific workflow, explaining each activity one by one; Section 4 illustrates the application of the workflow with a case study, and finally, Section 5 discuss possible improvements to the current implementation.

# 2   CADD, Molecular Docking, MD Simulation and Scientific Workflows

## 2.1   CADD, Molecular Docking and MD Simulation

New developments in structural and molecular biology and computer simulation tools over the past years have made possible a more accurate rational drug design (RDD) [9]. RDD involves a set of four steps [10]:

1. The target receptor (usually a protein) structure is analyzed using its 3D structure to identify probable binding sites;

2. Based on such probable binding sites, a set of possible ligands is selected and the receptor-ligand interactions can be tested and evaluated by simulations using a docking software;
3. The ligands that theoretically had the best interaction score to the receptor are selected, bought or synthesized, and then experimentally tested;
4. Based on the experimental results, a possible inhibitor is detected or the process returns to step 1.

Steps 1 and 2 constitute the CADD process. During the molecular docking (step 2), the ligand molecule assumes different orientations and conformations inside a defined binding pocket or region of the receptor and their interactions are systematically tested and evaluated (Figure 1a). A large number of evaluations has to be performed in order to identify the best ligand orientation and conformation inside the binding pocket. This information is computed in terms of the free energy of binding (FEB – the more negative, the more effective is the ligand-receptor association).

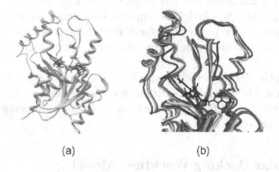

(a)                                   (b)

**Fig. 1.** The docking process. (a) The ligand molecule (in cyan and magenta) in two different orientations inside its InhA receptor (gray) binding pocket. (b) Superposition of five different *Mycobacterium tuberculosis* enzyme InhA conformations (cyan, yellow, magenta, red and green), generated by MD simulations [11], representing the flexibility of InhA bound to NADH (small molecule in blue). See section 3.1.

As ligands are usually small molecules, the different conformations they can assume inside the binding pocket are easily simulated by the docking software [4]. However, the limitation generally occurs when one wants to consider the receptor flexibility. There are a number of alternatives to incorporate at least part of the receptor mobility, but the use of many receptor structures has been characterized as the best alternative [5]. Therefore, one way to simulate the receptor flexibility is to use an ensemble of receptor conformations generated by MD simulations [12].

According to Sali [13], the studies of biological systems were initially limited to observation and interpretation of experimental data. The evolution of experimental techniques has allowed a deeper view of the biological processes by accessing the structural properties of biological macromolecules. These properties, in turn, can be deeply investigated by using the MD simulation methodology which simulates the molecular natural movements of biological molecules in atomic detail [14]. The result of a MD simulation is a series of instantaneous conformations or snapshots denominated the MD simulation trajectory. The InhA enzyme [15] from *Mycobacterium tuberculosis* has been shown to be considerably flexible [11] and was chosen to be our model of receptor (Figure 1b).

## 2.2  Scientific Workflows

According to the Workflow Management Coalition (WFMC) [16], workflow is "the automation of a business process, in whole or part, during which documents, information or tasks are passed from one participant to another for action, according to a set of procedural rules". Although this definition refers to "business process", workflow is not only employed by business applications. Wainer et al. [3] state that workflows can be also classified as ad-hoc workflows and scientific workflows.

Scientific workflows generally gather and merge data from various experiments, generating data from a computer model or performing data statistical analysis. In addition, scientific workflows can not be completely defined before it starts. While some tasks are being executed, one has to decide the next steps after evaluating the previous one [17]. Thus, the lack of complete knowledge about the processes in scientific applications has implications on modeling scientific workflows. The main assumption is that the models are inherently incomplete or change at any time [18].

As the molecular docking with respect to receptor flexibility is a scientific application, composed by a number of different software tools, in the present work we employed a scientific workflow management system to integrate modeling and execution of docking processes.

## 3  The Molecular Docking Workflow Model

Before the development of this scientific workflow all of the work had to be manually performed with FORTRAN computer programs and shell scripts. We developed our scientific workflow to model and run the docking processes considering the receptor flexibility explicitly, which is not a trivial task in ligand-receptor docking experiments [5]. The complexity of the receptor is usually the limiting issue. Receptors contain far more atoms than ligands, and therefore a very large number of degrees of freedom must be taken in account.

We adopted the Kua et al. [19] alternative approach to consider the receptor flexibility in docking experiments: to perform a series of dockings using, in each one of them, one different receptor snapshot. In our work, the receptor snapshots were generated by MD simulations [11] with the AMBER6.0 [8] package.

The flowchart of the developed workflow model is schematically shown in Figure 2. The activities in dashed lines correspond to those executed by the user, whilst the ones in solid lines are executed by the system without user intervention. The JAWE [6] and Shark [7] software tools were used to model and execute the workflow, respectively. These software tools are free, and can be used in the Linux environment (as well as AMBER6.0 and AutoDock3.05).

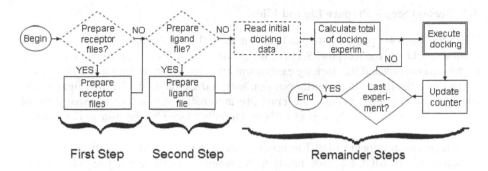

**Fig. 2.** Flowchart of the proposed molecular docking workflow model

## 3.1  Step Zero – MD Simulation of the Receptor

The first action in a molecular docking considering the explicit receptor flexibility is the execution of a MD simulation of the receptor, from which a series of receptor snapshots is generated. This step is called step zero because it is not modeled in Figure 2, since it is performed only once for each receptor.

The InhA enzyme from *Mycobacterium tuberculosis* has been shown to be considerably flexible [11] and was chosen to be our model of receptor (Figure 1b) in this work. Its explicit flexibility was obtained from a fully solvated MD simulation trajectory generated by the SANDER module of AMBER6.0 [8] as previously described [11]. MD data were collected for 3,100 ps (1.0 ps = $10^{-12}$s) and instantaneous snapshots were saved at every 0.5 ps, in files of 50 ps each. A total of 6,200 receptor snapshots were generated.

## 3.2  First Step – Prepare Receptor Files

This step has two parts: execution of PTRAJ and selection of snapshots relevant to the docking process. This step may not be necessary if the docking experiment only employs a single receptor.

PTRAJ is a AMBER6.0 utility that converts the trajectory of receptor snapshots generated by MD simulation into the PDB format [20]. A computer program was developed to establish the communication between Shark and PTRAJ. During workflow execution, the user must inform some parameters, such as the number of receptor amino acid residues and the first and last snapshots to be considered. These data, used as input parameters for PTRAJ, are stored into a workspace to be used during workflow execution. Thus, the user can easily change any input parameters and there is no need to modify individual scripts.

During the MD simulation, instantaneous snapshots were recorded at every 0.5 ps. Consequently, after 3,100 ps of MD simulation, 6,200 snapshots are generated and, hence, 6,200 PDB files are generated by PTRAJ. However, consecutive snapshots have closely related conformations. In order to trim down the redundancy of conformations we picked up snapshots separated by time intervals larger than 0.5ps. In our case study (see Section 4) we chose 1ps as the time interval to select snapshots and a total of 3100 snapshots were used in the docking experiments.

### 3.3  Second Step – Prepare Ligand File

This step has two parts. First, the ligand is placed in its initial position within the binding pocket of the receptor. Second, the proper ligand file is generated. This step is performed only once if the docking experiment employs the same ligand-receptor pair.

To place the ligand in its initial position within the binding pocket, the ligand PDB file and the receptor's average structure are automatically opened, by Shark, in the SwissPDBViewer [21]. The ligand is then manually placed by the user in the receptor structure.

Afterwards, the ligand PDB file needs to be transformed into a PDBQ format [4]. A ligand file in MOL2 format needs to be supplied as the input file. This can be basically done in two ways: through proprietary software such as MOE [22], or by downloading a MOL2 file from freely available public databases of small molecules. We developed a computer program that uses such a pre-existing MOL2 file and replaces its ligand coordinates for those correctly positioned in the receptor binding pocket. The module deftors of AutoDock3.05 software is then used to generate the PDBQ file from the MOL2 file.

### 3.4  Remainder Steps – Execution of the Docking Experiments

After the preparation of the receptor and ligand files, the docking experiments can be executed (Remainder Steps in Figure 2.). The flowchart in Figure 3 details the **Execute docking** subflow from Figure 2.

The docking experiments can be executed using the whole MD trajectory or only part of it. In the workflow the user is asked to inform the initial and final snapshots to be considered and a counter value, which indicates the next experiment to be performed (the counter value was introduced to prevent a restart from the beginning in case of workflow execution failure). As made for ligand and receptor files preparation, computer programs and shell scripts were developed to establish communication between the workflow and each of the AutoDock3.05 modules (Addsol, Mkgpf3, Mkdpf3, Autogrid and Autodock).

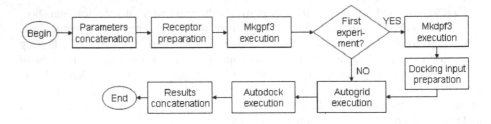

**Fig. 3.** Flowchart of the subflow Execute docking that executes the docking experiments

In Figure 3, the activity **Parameters concatenation** concatenates the counter value with execution file names from AutoDock3.05. The activity **Receptor preparation** generates the receptor in the PDBQS format using Addsol. **Mkgpf3 execution** generates the "Input.gpf" file, which contains the input parameters for Autogrid execution. **Mkdpf3 execution**, when executed, generates the Autodock parameters file "Input.dpf". Then, in **Docking input preparation**, "Input.dpf" is edited within a text editor and the user can easily modify the docking parameters. Subsequent executions of the subflow do not need to perform these activities.

The activity **Autogrid execution** executes the Autogrid module where the grid maps are defined for each ligand atom type. During **Autodock execution** the module Autodock is executed to calculate an estimate of the interaction between ligand and receptor in terms of the free energy of binding (FEB). At the end of the docking run, an output file is generated. This file contains information about all the tested ligand conformations, and the results are organized according to the best final docked energy (FEB) and the ligand root mean square deviation (RMSD) from the initial position.

The last activity, **Results concatenation**, collect the current docking energies (FEB) and RMSDs results and stores them in a results' list. An excerpt of this list is shown in Table 1, where each line represents the results of one docking experiment. This activity also compresses the Autodock output (to save disk space) and deletes the unnecessary files. All computer programs and scientific workflows for flexible receptor docking experiments were executed on Pentium III PCs of 1GHz and 256 MB RAM.

**Table 1.** Example of a list of results for the flexible InhA receptor-IPCF docking

| Time (ps) | Snapshot | RMSD (Å) | FEB (Kcal/mol) | Autogrid Execution Time (min.) | Autodock Execution Time (min.) |
|---|---|---|---|---|---|
| 1 | 2 | 6.3 | -9.9 | 4:50.02 | 10:22.50 |
| 2 | 4 | 6.2 | -10.2 | 4:06.81 | 10:08.61 |
| ... | ... | ... | ... | ... | ... |
| 3099 | 6198 | 3.9 | -9.9 | 4:39.90 | 9:41.25 |
| 3100 | 6200 | 4.0 | -9.7 | 4:30.58 | 10:00.94 |

## 4   Case Study

The validation of the proposed scientific workflow was carried out by performing docking experiments with the *Mycobacterium tuberculosis* enzyme InhA as the receptor and one large InhA ligand  (NADH [15]), and two small ones,  IPCF [23] and TCL [24].

### 4.1  The *M. Tuberculosis* Enzyme InhA

The InhA enzyme from *Mycobacterium tuberculosis* is the bona-fide target for one of the most important drugs (isoniazid) used in tuberculosis treatment. It was shown that this enzyme needs a NADH molecule as a cofactor for enzymatic activity. The activated isoniazid binds to the NADH molecule inside the receptor binding pocket to inhibit its activity [25], leading to mycobacterial death. It was shown that IPCF [23] and TCL [24] also interact with InhA, inhibiting its activity.

Knowing that the InhA enzyme constitutes a flexible receptor [11] and a ligand should bind to it in more than one enzyme conformation, and in order to understand these ligands affinities for the binding site, we developed a scientific workflow that automates the fully flexible molecular docking study to identify the characteristics of those ligand-InhA associations (see Section 3). The ligands molecules (NADH, IPCF and TCL) were docked in a number of different InhA (receptor) conformations previously generated by MD simulations [11].

## 4.2 Experiments

We performed three experiments, one for each ligand, to validate our implementation of the proposed workflow. The three ligands used in the experiments are shown in Figure 4.

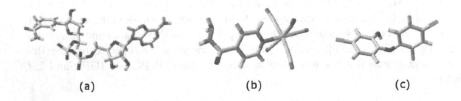

(a)                              (b)                              (c)

**Fig. 4.** Stick models of the three-dimensional structure of three InhA ligands: (a) NADH, (b) IPCF and (c) TCL. The ligands atoms are colored by type: Carbon (gray), Nitrogen (blue), Oxygen (red), Hydrogen (cyan), Phosphorus (yellow), Iron (orange), and Chlorine (green).

The same MD simulation trajectory of the receptor was used in all three experiments. Ligand docking to each of the 3,100 receptor snapshots was performed by the simulated annealing protocol including 10 runs with 100 cycles each, a total of 25,000 steps accepted or rejected, with selection of the ligand conformation presenting the minimum FEB. All docking processes and results concatenation were performed, with no human intervention, by the developed scientific workflow. The docking results are reported as in Table 2.

### 4.2.1 Docking Experiments with the NADH Ligand
After the receptor files preparation from the MD simulation snapshots, the NADH molecule, a large ligand (Figure 4a) presenting 52 atoms was generated through user interaction with the scientific workflow described above. The ligand was initially placed inside the receptor binding pocket, and its PDBQ file was prepared. As all receptor snapshots were superimposed, the initial ligand position, and therefore its PDBQ file, was the same for all 3,100 receptor snapshots considered.

### 4.2.2 Docking Experiment with the IPCF Ligand
The IPCF ligand molecule, containing 28 atoms, was prepared and initially placed in the receptor binding pocket as described for the NADH ligand in Section 4.2.1. Furthermore, the execution of the first step of the scientific workflow was not necessary since the receptor docking files had already been generated for the NADH docking.

### 4.2.3  Docking Experiment with the TCL Ligand
The TCL ligand molecule, with 24 atoms, was prepared as described in Section 4.2.2.

### 4.3  Results

The results summarized in Table 2 illustrate the efficiency of the proposed workflow and implementation of the automated molecular docking process.

**Table 2.** Results of the automated molecular docking process with flexible InhA receptor-ligands

| Ligand | Average FEB (-) (kcal/mol) | Total Number of FEB (-) | Total Numberof FEB (+) | Total Not Docked | Average RMSD (Å) |
|--------|-----------|-----------|-----------|-----------|-----------|
| NADH | -12.9 ± 4.2 | 2822 | 278 | 0 | 5.3 ± 2.2 |
| IPCF | -9.9 ± 0.6 | 3041 | 0 | 59 | 5.0 ± 1.4 |
| TCL | -8.8 ± 0.3 | 2890 | 0 | 210 | 6.9 ±1.9 |

Analysis of the NADH ligand docking output shows that not every receptor conformation leads to a good ligand association. From a total of 3,100 docking experiments, 278 presented a positive FEB (Table 2), i.e., a non favorable ligand-receptor interaction. On the other hand, this experiment has a good average FEB (-12.9 ± 4.2 kcal/mol) and an acceptable RMSD value, indicating that the NADH ligand remains inside its binding pocket during almost all of the docking process. The larger energy standard deviation, compared to the other tested ligands can be attributed to the ligand size. It is not easy to fit a bigger ligand in a binding pocket without observing some unfavorable interactions at some part of the ligand molecule.

The IPCF ligand docking to the receptor snapshots presents a good average FEB with low standard deviation (-9.9 ± 0.6 kcal/mol), indicating that this ligand binds to the receptor in almost all of its tested conformations. From a total of 3,100 docking experiments, only 59 did not converge (Table 2).

Analysis of the TCL ligand docking output list shows a good average FEB with low standard deviation (-8,8 ± 0.3 kcal/mol). The average RMSDs (Table 2) are higher than for the other ligands. This can be explained by the fact that, often the TCL ligand occupies the place originally occupied by the NADH (a natural ligand of the InhA receptor). As with the IPCF ligand, 210 docking experiments did not converge to a favorable interaction.

As expected, the InhA natural ligand, the NADH molecule, lead to the best docking results (lowest FEB), indicating that the receptor has a better affinity for this ligand. However, both IPCF and TCL ligands can also be considered as good ligands, as indicated by their FEB values.

The possibility of an automated docking process with explicit inclusion of the receptor flexibility can aid in a more realistic representation of the receptor-ligand interactions, and therefore improve the CADD process. Thus, the results above validate the developed workflow, turning the molecular docking process with explicit consideration of the receptor flexibility easier and more flexible to be executed with different receptor-ligand molecules.

## 5  Final Considerations

In this article we proposed to model and automate the molecular docking process with explicit consideration of the receptor flexibility. In order to achieve that, we developed a scientific workflow using JAWE and Shark software tools to model and execute the flexible docking process, respectively. To illustrate the efficiency of the proposed scientific workflow and its implementation we performed three molecular docking experiments, using a MD simulation trajectory of the *Mycobacterium tuberculosis* enzyme InhA as a model for the receptor flexibility and three InhA ligands: NADH, IPCF and TCL. These experiments showed that the scientific workflow efficiently executes all processes in an automated way, making it very easy to be executed for different flexible receptor-ligand molecules and by different users.

As future work we intend to introduce in the workflow a procedure to select receptor snapshots based on its binding energy (FEB) to a particular class of ligands. Thus, it is not going to be necessary to execute the experiment for all receptor's snapshots, saving processing efforts and, consequently, elapsed time.

## Acknowledgements

We thank the reviewers for helpful comments on the article. This project was supported by grants from FAPERGS and CNPq to ONS. ONS is a CNPq Research Fellow. KSM was supported by a CAPES M.Sc. scholarship.

## References

[1] Luscombe, N.M., Greenbaum, D., Gerstein, M.: What is Bioinformatics? A Proposed Definition and Overview of the Field. Meth. Inform. Med. 4, 346–358 (2001)

[2] Chagoyen, M., Kurul, M.E., De-Alarcón, P.A., Carazo, J.M., Gupta, A.: Designing and Executing Scientific Workflows with a programmable integrator. Bioinformatics 20, 2092–2100 (2004)

[3] Wainer, J., Weske, G.V., Medeiros, C.B.: Scientific Workflow Systems. In: Proceedings of the NFS Workshop on Workflow and Process Automation in Information Systems: State-of-the-art and Future Directions, Athens, Georgia, USA (1996)

[4] Goodsell, D.S., Olson, A.J.: Automated docking of substrates to proteins by simulated annealing. Proteins 8, 195–202 (1990)

[5] Carlson, H.A.: Protein flexibility is an important component of structure-based drug discovery. Curr. Pharm. Des. 8, 1571–1578 (2002)

[6] Mehta, N., Barter, R.H.: Design Document for JAWE2Openflow Project 2004 (accessed in December 2005), available in http://forge.objectweb.org/projects/jawe/

[7] Enhydra Shark (accessed in December 2005), available in http://forge.objectweb.org/projects/shark/

[8] Case, D.A., Pearlman, D.A., Caldwell, J.W., Cheathem III, T.E., Ross, W.R., Simmerling, C.L., Darden, T.A., Merz, K.M., Stanton, R.V., Cheng, A.L., Vincent, J.J., Crowley, M., Tsui, V., Radmer, R.J., Duan, Y., Pitera, J., Massova, I., Seibel, G.L., Singh, U.C., Weiner, P.K., Kollman, P.A.: AMBER 6.0. University of California, San Francisco (1999)

[9] Drews, J.: Drug discovery: A historical perspective computational methods for biomolecular docking. Curr. Opin. Struct. Biol., Elsevier Science 6, 402–406 (1996)

[10] Kuntz, I.D.: Structure-based strategies for drug design and discovery. Science 257, 1078–1082 (1992)

[11] Schroeder, E.K., Basso, L.A., Santos, D.S., Norberto de Souza, O.: Molecular Dynamics Simulation Studies of the Wild-Type, I21V, and I16T Mutants of Isoniazid-Resistant Mycobacterium tuberculosis Enoyl Reductase (InhA) in Complex with NADH: Toward the Understanding of NADH-InhA Different Affinities. Biophys. J. 89, 876–884 (2005)

[12] Lin, J-H., Perryman, A.L., Schames, J.R., McCammon, J.A.: Computational drug design accommodating receptor flexibility: the relaxed complex scheme. J. Am. Chem. Soc. 124, 5632–5633 (2002)

[13] Sali, A.: 100.000 Protein Structures for the Biologist. Nat. Struct..Biol. 5, 1029–1032 (1998)

[14] van Gunsteren, W.F., Berendsen, H.J.C.: Computer Simulation of Molecular Dynamics Methodology, Applications and Perspectives in Chemistry. Angew. Chem. Int. Ed. Engl. 29, 992–1023 (1990)

[15] Dessen, A., Quémard, A., Blanchard, J.S., Jacobs Jr., W.R., Sacchettini, J.C.: Crystal structure and function of the isoniazid target of Mycobacterium tuberculosis. Science 267, 1638–1641 (1995)

[16] Workflow Management Coalition – Terminology & Glossary: Document number WFMC-TC-1011. Document Status- Issue 3.0 (1999) (accessed in March 2006), available in http://www.wfmc.org/standards/docs/TC-1011_term_glossary_v3.pdf

[17] Ludäscher, B., Altintas, I., Berkley, C., Higgins, D., Jaeger, E., Jones, M.A., Lee, J., Tao, Y., Zhao, Y.: Scientific Workflow Management and the Kepler System. Concurrency and Computat.: Pract. Exper. 18, 1039–1065 (2005)

[18] Weske, M., Vossen, G., Medeiros, C.: Scientific Workflow Management: WASA Architecture and Applications. In: Revell, N., Tjoa, A.M. (eds.) DEXA 1995. LNCS, vol. 978, Springer, Heidelberg (1995)

[19] Kua, J., Zhang, Y., McCammon, A.: Studying Enzime Binding Specificity in Acetylcholinesterase Using a Combined Molecular Dynamics and Multiple Docking Approach. J. Am. Chem. Soc. 124, 8260–8267 (2002)

[20] Berman, H.M., Westbrook, J., Feng, Z., Gilliland, G., Bhat, T.N., Weissig, H., Shindyalov, I.N., Bourne, P.E.: PDB - Protein Data Bank. Nucl. Acids Res.. 28, 235–242 (2000)

[21] Guex, N., Peitsch, M.C.: SWISS-MODEL and the Swiss-PdbViewer: An environment for comparative protein modeling. Electrophoresis 18, 2714–2723 (1997)

[22] Chemical Computing Group, Inc. Montreal, Quebec, Canada. Molecular Operating Environment (MOE 2004.03) (accessed in July 2006), available in http://www.chemcomp.com

[23] Oliveira, J.S., Sousa, E.H.S., Basso, L.A., Palaci, M., Dietze, R., Santos, D.S., Moreira, I.S.: An inorganic iron complex that inhibits wild-type and an isoniazid-resistant mutant 2-trans-enoyl-ACP (CoA) reductase from Mycobacterium tuberculosis. Chem. Comm. 3, 312–313 (2004)

[24] Kuo, M.R., Morbidoni, H.R., Alland, D., Sneddon, S.F., Gourlie, B.B., Staveski, M.M., Leonard, M., Gregory, J.S., Janjigian, A.D., Yee, C., Musser, J.M., Kreiswirth, B., Iwamoto, H., Perozzo, R., Jacobs Jr., W.R., Sacchettini, J.C., Fodock, D.A.: Targeting tuberculosis and malaria through inhibition of enoyl reductase: compound activity and structural data. J. Biol. Chem. 278, 20851–20859 (2003)

[25] Rozwarski, D.A., Grant, G.A., Barton, D.H., Jacobs Jr., W.R., Sacchettini, J.C.: Modification of the NADH of the isoniazid target (InhA) from Mycobacterium tuberculosis. Science 279, 98–102 (1998)

# Gene Set Enrichment Analysis Using Non-parametric Scores

Ariel E. Bayá[1,*], Mónica G. Larese[1], Pablo M. Granitto[1],
Juan Carlos Gómez[1,2], and Elizabeth Tapia[1,3]

[1] Intelligent Systems Group, Instituto de Física Rosario, CONICET,
Bv. 27 de Febrero 210 Bis, 2000 Rosario, Argentina
[2] Laboratory for System Dynamics and Signal Processing, FCEIA, UNR,
Riobamba 245 Bis, 2000 Rosario, Argentina
[3] Communications Department, FCEIA, UNR, Riobamba 245 Bis, 2000 Rosario,
Argentina
{abaya,mlarese,granitto}@ifir.edu.ar
jcgomez@fceia.unr.edu.ar
etapia@eie.fceia.unr.edu.ar
http://www.ifir.edu.ar/

**Abstract.** Gene Set Enrichment Analysis (GSEA) is a well-known technique used for studying groups of functionally related genes and their correlation with phenotype. This method creates a ranked list of genes, which is used to calculate an enrichment score. In this work, we introduce two different metrics for gene ranking in GSEA, namely the Wilcoxon and the Baumgartner-Weiß-Schindler tests. The advantage of these metrics is that they do not assume any particular distribution on the data. We compared them with the signal-to-noise ratio metric originally proposed by the developers of GSEA on a type 2 diabetes mellitus (DM2) database. Statistical significance is evaluated by means of false discovery rate and $p$-value calculations. Results show that the Baumgartner-Weiß-Schindler test detects more pathways with statistical significance. One of them could be related to DM2, according to the literature, but further research is needed.

**Keywords:** GSEA, gene ranking, non-parametric statistical tests, statistical significance, DNA microarrays.

## 1 Introduction

DNA microarray technology has become a powerful technique that allows researchers to investigate the behaviour of thousands of genes simultaneously. After processing the information provided by the microchip, it is possible to study the correlations between the expression or suppression of genes and the consequent occurrence of certain diseases.

Recent studies suggest that single gene analysis could lead to non-significant statistical interpretations [5]. Some limitations of this approach are discussed

---

* Author to whom all correspondence should be addressed.

M.-F. Sagot and M.E.M.T. Walter (Eds.): BSB 2007, LNBI 4643, pp. 12–21, 2007.
© Springer-Verlag Berlin Heidelberg 2007

in [13]. In order to overcome this, the authors in [5,13] propose a new method called Gene Set Enrichment Analysis (GSEA), which analyzes co-acting genes in the same metabolic pathway.

GSEA makes use of a scoring metric to rank all genes from the microarray into a unique list. After that, a running sum, named Enrichment Score ($ES$), is computed for each pathway. The original metric proposed in [5] is the Signal to Noise Ratio ($SNR$), which assumes that the probe samples for each gene are normally distributed. This is a limitation in the sense that it is very common to find non-Gaussian distributions in microarray experiments. In addition, this scoring metric is very sensitive to sample outliers which are also very common in microarray data. However, GSEA is versatile enough to incorporate other scoring metrics.

In the present work, we propose the use of GSEA in combination with two metrics based on non-parametric tests to calculate the scores and rank the genes, namely the Wilcoxon [10] and the Baumgartner-Weiß-Schindler [1] tests. These metrics provide a score based on ranking tests and no assumptions about the distributions are made. Therefore they are not affected by the limitations described above. These metrics have already been used in [7], in the context of detection of differentially expressed genes from microarray data. To the best of the authors' knowledge, the combination of GSEA and non-parametric scoring metrics has not been proposed in the literature before for the purposes of enrichment analysis of gene sets.

The computations involved in this work were performed resorting to a Java based software package developed by the authors. Several databases supported at NCBI[1] were also used for consulting.

The rest of the paper is organized as follows. In section 2 the GSEA method is briefly described. The different scoring metrics used in the paper are presented in section 3. The statistical significance procedure implemented in this work is detailed in section 4. In section 5 the dataset used in the experiments is described. The results are presented and discussed in section 6. Finally, some conclusions are drawn in section 7.

## 2   Gene Set Enrichment Analysis (GSEA)

GSEA is a method developed to analyze the statistical significance and the correlation of a gene set with the phenotype. This idea was introduced by Mootha *et al.* in 2003 [5] and then improved and tested on different datasets by Subramanian *et al.* in 2005 [13], where it is described in more detail. In the present work the latter approach was implemented and extended to include the use of non-parametric scoring tests.

First, GSEA ranks all $N$ genes in a global list using a scoring metric which expresses the correlation between each gene $g_i$ and the phenotype (see section 3).

---

[1] NCBI: National Center for Biotechnology Information, Bethesda, MD 20894 (http://www.ncbi.nlm.nih.gov/)

Given a gene set $G$ of size $N_G$, the $ES$ for $G$ is calculated as

$$ES(G) = \max_{1 \leq i \leq N} |P_h(G,i) - P_m(G,i)| \times \text{sign}\left(\max_{1 \leq i \leq N} P_h(G,i) - P_m(G,i)\right) \quad (1)$$

where

$$P_h(G,i) = \sum_{g_i \in G, j \leq i} \left(\frac{|m_j|^w}{\sum_{g_i \in G} |m_j|^w}\right) \quad (2)$$

is the fraction of genes in $G$ up to position $i$ in the ranked list ("hits") weighted by their correlation, and

$$P_m(G,i) = \sum_{g_i \in G, j \leq i} \frac{1}{N - N_G} \quad (3)$$

is the fraction of genes not in $G$ up to position $i$ in the ranked list ("misses"). Here, $m_j$ represents the correlation of $g_i$ with the phenotype and $w$ is a constant.

The $ES(G)$ in (1) is then the maximum deviation from zero of the difference between $P_h(G)$ and $P_m(G)$. If $w \neq 0$, the sums in (1) are weighted by $m_j$ to the power $w$. In this case the genes that are located at the top or bottom of the ranked list are enhanced, while those others positioned in the middle (low correlation with the phenotype) have a weaker contribution to the $ES(G)$.

## 3   Scoring Metrics

Scoring metrics are used by GSEA to rank the genes into a global list according to their correlation with the phenotype's class. The original metric proposed by GSEA is the $SNR$ (see section 3.1). In the present work we propose two other alternatives, namely Wilcoxon and Baumgartner-Weiß-Schindler tests, which are described in sections 3.2 and 3.3, respectively.

### 3.1   Signal-to-Noise-Ratio ($SNR$)

In the context of scoring genes, the $SNR$ is defined as the difference in means of the probe samples for each phenotype's class divided by the sum of their standard deviations [2,11],

$$SNR = \frac{(\mu_0 - \mu_1)}{(\sigma_0 + \sigma_1)}. \quad (4)$$

Manoli et al. [4] introduced a variation in the score metric by computing the absolute value of the numerator in (4),

$$|SNR| = \frac{|(\mu_0 - \mu_1)|}{(\sigma_0 + \sigma_1)}. \quad (5)$$

The use of absolute values causes both genes having the highest positive and negative correlations with phenotype to be located at the top of the ranked list.

## 3.2  Wilcoxon Rank Sum Test

The Wilcoxon Rank Sum non-parametric test (hereafter denoted as $W$ test) is applicable when data coming from two independent samples can be converted to ordinal ranks. Despite the fact that this conversion causes some loss of information, this method has the strength that does not make any assumption about the probability distribution from which the samples are taken.

Very commonly, microarray experiments have a large number of outliers and normality assumptions can not be made [3]. The $W$ test can still be applied under these conditions, where the $t$-test is not applicable because of its strong parametric assumptions. The reader is referred to [10] for a more extensive description of this test.

The method makes the assumption that the underlying sample distributions have the same shape (not necessarily normal), except for the median value. The null hypothesis states that the medians from the two populations are the same, while the alternative hypothesis states that they are different.

The score is calculated for each gene by arranging the expression values of all probes from both phenotype classes in non-decreasing order. Then, each one of the $N_p$ probes is assigned a rank $(R^i)$ from 1 to $N_p$. The Wilcoxon statistic is computed next as the sum of the $N_0$ ranks for the first class, as shown in (6).

$$W = \sum_{i=1}^{N_0} R^i. \tag{6}$$

## 3.3  Baumgartner-Weiß-Schindler Test

The Baumgartner-Weiß-Schindler non-parametric test (hereafter denoted as $BWS$ test) makes the same assumptions about the samples as the $W$ test does. However, it has probed to be less conservative and to yield better results. [6] The reader is referred to [1] for further details about this test. Neuhäuser and Senske [7] successfully applied this test for the detection of differentially expressed genes.

The distribution functions underlying the two samples are assumed to be identical except for a shift in their locations, *i.e.*: $F(x) = G(x - \theta)$, for every $x$ and $-\infty < \theta < +\infty$. The null hypothesis states that $\theta$ is null. The alternative hypothesis states that $\theta$ is not null.

The $BWS$ statistic is computed as follows

$$BWS = \frac{(B_0 + B_1)}{2}, \tag{7}$$

with

$$B_0 = \frac{1}{N_0} \sum_{i=1}^{N_0} \frac{(R_0^i - \frac{(N_1+N_0)}{N_0}\cdot i)^2}{\frac{i}{(N_0+1)}\cdot(1 - \frac{i}{(N_0+1)})\cdot\frac{N_1\cdot(N_1+N_0)}{N_0}}, \tag{8}$$

$$B_1 = \frac{1}{N_1} \sum_{j=1}^{N_1} \frac{(R_1^j - \frac{(N_1+N_0)}{N_1}\cdot j)^2}{\frac{j}{(N_1+1)}\cdot(1 - \frac{j}{(N_1+1)})\cdot\frac{N_0\cdot(N_1+N_0)}{N_1}}, \tag{9}$$

where $N_0$ and $N_1$ are the sample sizes for classes 0 and 1, respectively, and $R_0^i$ and $R_1^j$ are the Wilcoxon ranks for the $i$-th and $j$-th probes for classes 0 and 1, respectively.

## 4    Statistical Significance Analysis

The GSEA method is implemented using the four scoring metrics described in section 3, i.e., $SNR$, $|SNR|$, $W$ test and $BWS$ test. As a result four different values of the $ES$ for each gene set $G$ are computed. The statistical significance is then assessed by means of the nominal $p$-value calculation and multiple hypothesis testing, as described in [13].

A randomization test was implemented using 1000 uniformly distributed random permutations of the class labels and recalculating the $ES$ values each time, in order to generate a null distribution of the $ES$. The nominal $p$-values were then computed as the ratio between the observed $ES$ and the corresponding positive or negative part, depending on the sign of the $ES$, of the null distribution.

False Discovery Rate ($FDR$) calculation [12] was employed to assess multiple hypothesis testing. It was achieved by first normalizing the $ES(G)$ by the total size of the set obtaining the normalized enrichment score ($NES(G)$). The null distribution was also computed for the $NES(G)$. Then, the $FDR$ was calculated by comparing the tails of the observed $NES(G)$ and its null distribution. The $FDR$ value represents the probability for a given $NES(G)$ of being a false positive. The procedure is explained in more detail in [13].

## 5    Dataset Description

The normalized dataset, as well as the set of 149 curated pathways, which we used in this work is the one provided by [5].

A reduced version of the DNA microarray dataset is available for public download from the Broad Institute Website[2]. It consists of 34 transcriptional profiles from 17 patients affected with Type 2 Diabetes Mellitus (DM2) and 17 individuals with Normal Glucose Tolerance (NGT). The total number of genes is 22283.

The 149 gene sets are comprised by 113 groups of genes involved in metabolic pathways and 36 groups of GNF mouse expression clusters. Eleven out of the

---

[2] http://www.broad.mit.edu/gsea/resources/datasets_index.html

113 pathways were manually curated by Mootha and co-workers and reported in [5] and they correspond to: oxidative phosphorylation, glycolysis, glycogen metabolism, pyruvate metabolism, ketogenesis, reactive oxygen species homeostasis, insulin signaling, gluconeogenesis, free fatty acid metabolism, mitochondria, and Krebs cycle.

# 6   Discussion

The motivation of this work is to test different metrics in order to obtain new gene sets that could eventually be biologically related to the disease. GSEA was implemented using each one of the four scoring metrics described in section 3, computing the $ES$, $NES$, $p$-value and $FDR$-$q$-value for each one of the 149 pathways.

Figure 1 shows the behaviour of the newly introduced statistical metrics compared with the $SNR$ metric used by Mootha *et al.* [5].

The different plots in Figure 1 show the normalized index of the genes ordered by one of the newly tested metrics (*i.e.*: $|SNR|$, $W$ and $BWS$ scores) as a function of the normalized index of the same genes ordered by $SNR$. It is easy to see that all four metrics yield different rankings.

As noted by Subramanian *et al.* [13], there is a bias in the bimodal $ES$ distribution as a result of an unequal representation of the phenotype by the gene sets. For each plot in Figure 2 a null hypothesis distribution was generated by taking

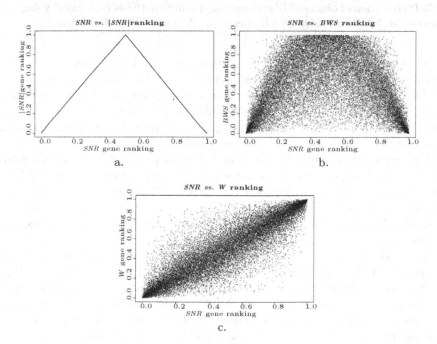

**Fig. 1.** Comparison of different metrics. **a.** $|SNR|$ metric *vs.* $SNR$ metric. **b.** $BWS$ metric *vs.* $SNR$ metric. **c.** $W$ metric *vs.* $SNR$ metric.

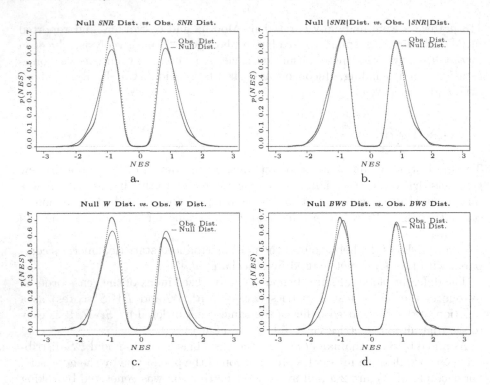

**Fig. 2.** Distribution of observed (*dotted line*) and null hypothesis (*solid line*) values for all metrics. **a.** $SNR$ metric. **b.** $|SNR|$ metric. **c.** $W$ metric. **d.** $BWS$ metric.

1000 random permutations of the phenotype using the C2 symbol database for pathways and the collapsed Diabetes database [13]. As expected, in all cases a bimodal distribution was obtained. Positive and negative values were separately normalized to correct the bias in the distribution, as well as the observed distribution. As it can be noticed from all plots for both distributions, null and alternative hypothesis have a very similar shape.

After performing the enrichment score calculation and the statistical significance analysis, the gene sets that simultaneously verify the constraints: i) $FDR$-$q$-value less than 0.25 and ii) $p$-value less than 0.05, were retained. We chose the standard 0.25 and 0.05 thresholds for the $FDR$-$q$-value and $p$-value, respectively, following [13,5]. The lists of significant pathways detected in each case are presented in Table 1 to Table 4.

It is noticeable that in the four situations the most significant gene set corresponds to the one related to oxidative phosphorylation (OXPHOS-HG-U133A-probes, hereafter denoted as OXPHOS), which is also the only significant pathway indicated by using the $SNR$, $|SNR|$ and $W$ scoring metrics. It agrees with the results obtained by Mootha *et al.* [5], who also obtained the same significant pathway, although not through $FDR$-$q$-value analysis.

**Table 1.** Pathways with $FDR$ $q$-value$< 0.25$ and nominal $p$-value$< 0.05$, using the signal-to-noise ratio scoring metric and a weight $w = 0$ in (2)

| SNR | | |
|---|---|---|
| Pathway | Nominal $p$-value | $FDR$ $q$-value |
| OXPHOS-HG-U133A-probes | 0.0043 | 0.0533 |

**Table 2.** Pathways with $FDR$ $q$-value$< 0.25$ and nominal $p$-value$< 0.05$, using the absolute signal-to-noise ratio scoring metric and a weight $w = 0$ in (2)

| $|SNR|$ | | |
|---|---|---|
| Pathway | Nominal $p$-value | $FDR$ $q$-value |
| OXPHOS-HG-U133A-probes | 0.0304 | 0.1933 |

**Table 3.** Pathways with $FDR$ $q$-value$< 0.25$ and nominal $p$-value$< 0.05$, using the Wilcoxon rank sum scoring metric and a weight $w = 0$ in (2)

| Wilcoxon Rank Sum | | |
|---|---|---|
| Pathway | Nominal $p$-value | $FDR$ $q$-value |
| OXPHOS-HG-U133A-probes | 0.0200 | 0.1467 |

**Table 4.** Pathways with $FDR$ $q$-value$< 0.25$ and nominal $p$-value$< 0.05$, using the Baumgartner-Weiß-Schindler scoring metric and a weight $w = 0$ in (2)

| Baumgartner-Weiß-Schindler metric | | |
|---|---|---|
| Pathway | Nominal $p$-value | $FDR$ $q$-value |
| OXPHOS-HG-U133A-probes | 0.0085 | 0.0485 |
| FA-HG-U133A-probes | 0.0020 | 0.1152 |
| mitochondr-HG-U133A-probes | 0.0102 | 0.1782 |
| human-mitoDB-6-2002-HG-U133A-probes | 0.0324 | 0.2090 |

In our work, however, the $BWS$ metric additionally finds three more significant gene sets. They correspond to FA-HG-U133A-probes, mitochondr-HG-U133A-probes and human-mitoDB-6-2002-HG-U133A-probes (see Table 4). The latter two are also indicated by Mootha *et al.* [5] to be the two gene sets with the highest enrichments after OXPHOS, even though in their work they do not label them as significant based on their statistical analysis. They notice that these pathways overlap OXPHOS, and thus their enrichment could be explained because of the genes they share. This is not the case for the FA-HG-U133A-probes, since this gene set does not overlap OXPHOS at all. It has partial overlap with mitochondr-HG-U133A-probes and human-mitoDB-6-2002-HG-U133A-probes (88.24% and 97.06%, respectively).

FA-HG-U133A-probes gene set was internally curated by Mootha and co-authors in [5] and is composed by genes related to free fatty acid metabolism.

In the literature, fatty acid metabolism has been considered importantly related to oxidative phosphorylation processes. Petersen *et al.* [8] conclude that insulin resistance in patients with type 2 diabetes may be related with dysregulation of intramyocellular fatty acid metabolism, probably caused by inherited defects in oxidative phosphorylation at the mitochondria. Rufer *et al.* [9] investigated the downregulation activity of the carnitine palmitoyltransferase (CPT) system in the development of novel drugs for diabetes treatment. CPT1A and CPT2 genes are involved in the fatty acid metabolism and are members of this pathway. Additionally, Wood [14] states that the presence of a large amount of fatty acids and a low fatty acid oxidation may cause insulin resistance and, eventually, type 2 diabetes mellitus.

The use of the GSEA method in combination with the $BWS$ metric resulted in the detection of three extra significant gene sets. The studies referenced above suggest the biological association between one of them, namely the free fatty acid metabolism pathway, with type 2 diabetes mellitus. The other two extra gene sets detected, mitochondr-HG-U133A-probes and human-mitoDB-6-2002-HG-U133A-probes could be explained on the basis of their overlap with OXPHOS and FA-HG-U133A-probes. However this is not the case for FA-HG-U133A-probes, which has no overlap with OXPHOS and is ranked second in the highest significant pathway list. Further research is needed to confirm the hypothesis of the link between the occurrence of type 2 diabetes and variations on the differential expression of the fatty acid metabolism gene set.

# 7  Concluding Remarks

In this paper the use of two non-parametric scoring metrics, namely Wilcoxon and Baumgartner-Weiß-Schindler tests, in combination with GSEA is proposed for the study of functionally related genes and their correlation with type 2 diabetes mellitus. The statistical analysis ($FDR$-$q$-value and $p$-value) with the different metrics agreed in detecting OXPHOS as the most significant pathway, confirming previous studies. Additionally, the use of the Baumgartner-Weiß-Schindler metric allowed to detect three extra gene sets apart from OXPHOS, as the next most significant pathways. In this sense this metric seems to be more powerful than the standard signal-to-noise-ratio. The second in significance level gene set (the free fatty acid metabolism group) has been indicated as possibly related to the disease in previous biological studies available from the literature. Further biological research is needed to experimentally confirm this association. The use of the Baumgartner-Weiß-Schindler metric in combination with GSEA seems to be a less conservative and more powerful technique for the purposes of identifying differentially expressed gene sets.

**Acknowledgements.** Work supported by CONICET, ANPCyT under grant PICT 11-15132 and Universidad Nacional de Rosario.

# References

1. Baumgartner, W., Weiß, P., Schindler, H.: A Nonparametric Test for the General Two-Sample Problem. Biometrics 54, 1129–1135 (1998)
2. Golub, T.R., Slonim, D.K., Tamayo, P., Huard, C., Gaasenbeek, M., Mesirov, J.P., Coller, H., Loh, M.L., Downing, J.R., Caligiuri, M.A., Bloomfield, C.D., Lander, E.S.: Molecular Classification of Cancer: Class Discovery and Class Prediction by Gene Expression Monitoring. Science 286, 531–537 (1999)
3. Liu, L., Hawkins, D.M., Ghosh, S., Young, S.S.: Robust Singular Value Decomposition Analysis of Microarray Data. PNAS 100(23), 13167–13172 (2003)
4. Manoli, T., Gretz, N., Gröne, H.-J., Kenzelmann, M., Eils, R., Brors, B.: Group Testing for Pathway Analysis Improves Comparability of Different Microarray Datasets. Bioinformatics 22(20), 2500–2506 (2006)
5. Mootha, V.K., Lindgren, C.M., Eriksson, K.-F., Subramanian, A., Sihag, S., Lehar, J., Puigserver, P., Carlsson, E., Ridderstråle, M., Laurila, E., Houstis, N., Daly, M.J., Patterson, N., Mesirov, J.P., Golub, T.R., Tamayo, P., Spiegelman, B., Lander, E.S., Hirschhorn, J.N., Altshuler, D., Groop, L.C.: PGC-1α-Responsive Genes Involved in Oxidative Phosphorylation are Coordinately Downregulated in Human Diabetes. Nature Genetics 34(3), 267–273 (2003)
6. Neuhäuser, M.: An Exact Two-Sample Test Based on the Baumgartner-Weiß-Schindler Statistic and a Modification of the Lepage's Test. Communications in Statistics-Theory and Methods 29(1), 67–78 (2000)
7. Neuhäuser, M., Senske, R.: The Baumgartner-Weiß-Schindler Test for the Detection of Differentially Expressed Genes in Replicated Microarray Experiments. Bioinformatics 20(18), 3553–3564 (2004)
8. Petersen, K.F., Dufour, S., Befroy, D., Garcia, R., Shulman, G.I.: Impaired Mitochondrial Activity in the Insulin-Resistant Offspring of Patients with Type 2 Diabetes. The New England Journal of Medicine 350, 664–671 (2004)
9. Rufer, A.C., Thoma, R., Benz, J., Stihle, M., Gsell, B., De Roo, E., Banner, D.W., Mueller, F., Chomienne, O., Hennig, M.: The Crystal Structure of Carnitine Palmitoyltransferase 2 and Implications for Diabetes Treatment. Structure 14, 713–723 (2006)
10. Sheskin, D.J.: Handbook of Parametric and Nonparametric Statistical Procedures, 2nd edn. Chapman & Hall/CRC, Boca Raton (2000)
11. Shipp, M.A., Ross, K.N., Tamayo, P., Weng, A.P., Kutok, J.L., Aguiar, R.C.T., Gaasenbeek, M., Angelo, M., Reich, M., Pinkus, G.S., Ray, T.S., Koval, M.A., Last, K.W., Norton, A., Lister, T.A., Mesirov, J., Neuberg, D.S., Lander, E.S., Aster, J.C., Golub, T.R.: Diffuse Large B-cell Lymphoma Outcome Prediction by Gene-Expression Profiling and Supervised Machine Learning. Nature Medicine 8(1), 68–74 (2002)
12. Storey, J.D., Tibshirani, R.: Statistical Significance for Genomewide Studies. PNAS 100(16), 9440–9445 (2003)
13. Subramanian, A., Tamayo, P., Mootha, V.K., Mukherjee, S., Ebert, B.L., Gillette, M.A., Paulovich, A., Pomeroy, S.L., Golub, T.R., Lander, E.S., Mesirov, J.P.: Gene Set Enrichment Analysis: A Knowledge-Based Approach for Interpreting Genome-Wide Expression Profiles. PNAS 102(43), 15545–15550 (2005)
14. Wood, P.A.: Genetically Modified Mouse Models for Disorders of Fatty Acid Metabolism: Pursuing the Nutrigenomics of Insulin Resistance and Type 2 Diabetes. Nutrition 20(1), 121–126 (2004)

# Comparison of Simple Encoding Schemes in GA's for the Motif Finding Problem: Preliminary Results

Giovanna Martínez-Arellano and Carlos A. Brizuela

Computer Sciences Department
CICESE Research Center
Km 107 Carr. Tijuana-Ensenada, Ensenada, B.C., México
{gimartin, cbrizuel}@cicese.mx
+52–646–175–0500

**Abstract.** The DNA motif finding problem is of great relevance in molecular biology. Weak signals that mark transcription factor binding sites involved in gene regulation are considered to be challenging to find. These signals (motifs) consist of a short string of unknown length that can be located anywhere in the gene promoter region. Therefore, the problem consists on discovering short, conserved sites in genomic DNA without knowing, *a priori*, the length nor the chemical composition of the site, turning the original problem into a combinatorial one, where computational tools can be applied to find the solution. Pevzner and Sze [7], studied a precise combinatorial formulation of this problem, called *the planted motif problem*, which is of particular interest because it is a challenging model for commonly used motif-finding algorithms [15]. In this work, we analyze two different encoding schemes for genetic algorithms to solve the planted motif finding problem. One representation encodes the initial position for the motif occurrences at each sequence, and the other encodes a candidate motif. We test the performance of both algorithms on a set of planted motif instances. Preliminary experimental results show a promising superior performance of the algorithm encoding the candidate motif over the more standard position based scheme.

## 1 Introduction

The Motif Finding Problem can be defined as to find short conserved sites in DNA sequences without knowing, *a priori*, the length nor the bases that compose them. Until now, algorithms have been developed to identify motifs that appear in several sequences. In this case, instead of looking for a single site in one DNA sequence, they look for several sites containing substrings that are very alike (some bases may change) taking several DNA sequences. There are two categories for this kind of algorithms. The first one, that can be called "multiple genes, one species", assumes that a single motif (degenerated) is embedded in some or all DNA sequences and finds a *consensus* and its occurrences. The second one, called "single gene multiple species", takes several orthologous DNA sequences (same gene in different species) and finds well conserved sites.

M.-F. Sagot and M.E.M.T. Walter (Eds.): BSB 2007, LNBI 4643, pp. 22–33, 2007.
© Springer-Verlag Berlin Heidelberg 2007

We can identify three cases of the motif finding problem [9]. In the first case, the motif appears in each of the input DNA sequences with some mutations. In the second case, the motif appears only in some sequences presenting some mutations, and, in the third case, more than one occurrence of the same motif can appear in a single sequence presenting different mutations. Most of the algorithms are designed to have a good performance for the first model, nevertheless, they are tested with the other two, showing, most of the time, a lower performance.

Some of the existing algorithms like Gibbs Sampler [11] and MEME [1] are used in practice to solve the motif finding problem, although, for planted motifs, Random Projections [4] and Pattern Branching [15], among others, get better results.

Evolutionary algorithms have also been proposed to deal with this problem. However, a comparison of the two most obvious encoding schemes has been scarce [21]. It will be interesting to know which scheme works better for which variants of the problem. The work we present here is a small step towards this direction.

The remainder of the paper is organized as follows. Section 2 states the problem we are going to solve and present a brief description of previous work on the subject. Section 3 describes the proposed algorithm along with the standard genetic algorithm. Section 4 discuss the experimental setup and presents the preliminary results. Section 5 presents the conclusions and some ideas for future research.

## 2  Problem Definition

To define the Motif Finding Problem, we will start by defining what a string is. A string $S$ is an ordered list of characters written contiguously from left to right. For any string $S$, $S[i..j]$ is the (contiguous) substring of $S$ that starts at position $i$ and ends at position $j$ of $S$ [7].

A motif is a substring $s$ of length $l$ that can or cannot appear in a given string $S$. An occurrence of a $(l, d)$-motif is a substring $s$ of length $l$ that differs in at most $d$ characters from the motif.

In the Planted Motif Finding Problem [14], each input string $S_i$ ($i \in \{1, 2, ..., N\}$) contains a planted occurrence of an $(l, d)$-motif, having an initial position $j \in \{1, 2, ..., T - l + 1\}$, where $T$ is the length of the string $S_i$ and $l$ the length of the motif and of its occurrences. We assume, without lost of generality, that all strings $S_i$ are of the same length.

The objective of the planted motif problem is to find all the occurrences of the $(l, d)$-motifs that appear in each of the $N$ input strings without knowing, a priori, the motif.

We will now formally define the Planted Motif Finding Problem as the following input/output requirements.

**Input:** A set of $N$ strings $\{S_1, S_2, ..., S_N\}$ each of length $T$ over the alphabet $\{a, c, g, t\}$, where each string contains a planted occurrence of the motif. The length of the motif ($l$) and the maximum number $d$ of possible mutations in its occurrences.

**Output:** A set $P$ of initial positions $\{p_1, p_2, ..., p_N\}$ that correspond to the first character of each of the $N$ occurrences that where found. Those $N$ occurrences of the inserted $(l, d)$-motif. Each $p_i \in \{1, 2, \cdots, T - l + 1\} \cup \{0\}$, with $p_i = 0$ means that sequence $S_i$ contains no occurrences of the motif.

Notice that this problem was already implicitly defined in [17]. Although the *planted motif problem* was not explicitly defined, the author introduced two variants of the motif finding problem and, one of them includes the motif finding problem as it is defined above.

The complexity of this problem and its variants (all NP-Hard [23]) has motivated the development of efficient heuristics to deal with them.

## 2.1   Previous Work

This problem has been dealt with different approaches. Some of the existing algorithms like CONSENSUS [8], Gibbs [11], and MEME [1] are local search based algorithms. Some others [2], [3], [19], use enumeration strategies. In [17] two algorithms based on the construction of suffix trees are proposed. Both time and space complexity bounds are improved in this paper comparing to Sagot et al. [18] and to the ones produced by Waterman et al. [22].

In [4] an algorithm based on random projections of the motif is proposed. Constructed under the planted motif model, this algorithm obtains a good performance compared with the commonly used methods like Gibbs [11] and MEME [1]. In [23] the Motif Finding Problem is defined as to find a local alignment of multiple sequences without gaps using the sum-of-pairs scoring scheme and the filogenetic distance. Combinatorial techniques like branch pruning and linear programming are used to solve the problem.

In [15] an algorithm based on the planted motif model, looking for the planted motif instead of its occurrences, is proposed. The main idea of this approach, known as Pattern Branching, is that if we have a real occurrence, then we can obtain the original motif by changing, in this occurrence, exactly $d$ bases. So if we analyze all the possible substrings in all DNA strings, we will have, at the end, the best motif that minimizes a given score. The second encoding we propose here is based on some ideas of this algorithm. Pattern Branching [15] obtains solutions that are comparable to the ones obtained by Random Projections [4], with a notorious improvement in the computation time.

Tracing back solutions in the evolutionary algorithms context, we found several approaches. The application of genetic algorithms for the motif finding problem was first introduced in [21]. Two algorithms were proposed. The first one, GA1, is based on the position weight matrix, where each chromosome represents an alignment of all sequences. The algorithm tries to find all alignments, which maximize the sum of the maximum frequencies of nucleotides at each position of the alignment over the motif length. In the second algorithm, GA2, the chromosome represents a candidate consensus string of length $l$ and the algorithm tries to maximize the score of this consensus string with respect to all sequences in the data set. Both algorithms were tested with synthetic data and some real samples (in the case of GA2). This paper concludes that GA2 is faster, because the search space is smaller, but no computation time results are provided nor compared with other state-of-the-art algorithms. Since GA2 has a higher probability of finding correct motifs, it is concluded that it is applicable for signal identification in real biological data. Basically, this work pretends to determine in general if genetic algorithms are suitables for the motif finding problem studying the probabilities of the algorithms to identify correct motifs. A similar study is presented in our present work, but our main goal is to determine which scheme, position based or motif based, works better for each of the motif finding problem variants. To begin in this direction, here we focus our experiments in the planted motif problem defined in [14] and compare results with Gibbs Sampler [11].

In [5], a standard GA is proposed to deal with the problem, here the representation is given by a vector of initial positions, corresponding to each occurrence of the motif. The results are compared with the ones produced by Gibbs Sampler [11], BioProspector [13], MEME [1], Consensus [8] and AlignACE [16] on a small set of real cases with competitive results. In [12] a more complicated approach is adopted, each individual is encoded as a set of candidate motif patterns generated at random, one motif pattern per sequence. The evaluation function for a single sequence is computed as the best matching percentage of all subsequences in that sequence, and the overall fitness score is the summation of individual fitness scores for all sequences. Results are compared with Gibbs [11] and MEME [1], showing a superior performance over them. In [10] GAMOT algorithm is proposed and applied to the motif planted problem. This algorithm makes a fast search of candidate motifs to take as initial population before the genetic algorithm begins. GAMOT is tested only with synthetic data and is compared with Random Projections, GA1, GA2 and some exhaustive search algorithms, achieving better results in computation time (when comparing to the exact methods) and quality of solutions and computation time when comparing to the other heuristics.

## 3   The Algorithms

The ultimate goal here is to see what level of performance can be achieved with the evolutionary approach. The first step in that direction is to compare the most obvious encoding schemes to see which is more appropriate for the problem. One encoding will be to define an individual as a vector of initial positions for candidate occurrences of the motif. This approach is similar to the one adopted in [5]. The other encoding will define a chromosome to be the motif itself, resembling other approaches which are not based on the evolutionary paradigm. In the following sections we explain in detail both encodings.

### 3.1   Position Based Encoding Genetic Algorithm: PbGA

To implement the algorithm, the representation of the individual and the genetic operators (crossover and mutation) must be defined. Since the solution to this

| a | g | g | c | t |
| a | g | t | c | t |
| c | g | g | c | t |
| a | g | c | c | g |
| a | g | t | a | t |

**Fig. 1.** Aligning of five occurrences of $(l, d)$-motifs given by individual $\mathbf{i} = [3, 29, 147, 12, 87]$

problem is given by a set of initial positions for each motif occurrence on each string $S_i$, we will define our individual as $\mathbf{i} = [p_1, p_2, ..., p_N]$, where $p_j$ represents the initial position of the occurrence found in the string $S_j$. The fitness of the individual is calculated aligning the $N$ occurrences it represents and adding for each column the character that is repeated the most. Let us assume we are looking for a motif of length five ($l = 5$). Now we will compute the fitness of an example individual given by $\mathbf{i} = [3, 29, 147, 12, 87]$. To do this we search for the length $l = 5$ string starting at position 3 in $S_1$, the one starting at position 29 in $S_2$, and so on. Let us assume that after aligning all length $l(= 5)$ substrings defined by individual $\mathbf{i}$ we have the result shown in Figure 1.

We see that, in the first column, character $a$ has the largest number of occurrences (4), so the score for this column is 4. Following the same procedure we obtain the total score as $Score(\mathbf{i}) = 4 + 5 + 2 + 4 + 4 = 19$.

In this case, we will maximize the score (which we adopted as fitness) of the individuals through the generations. It is not hard to see that the cost for computing this score is $O(l * N)$. Notice that the fitness used in [5] is different but considers the same information, i.e. the number of characters occurrences in each column. Once the individual representation is defined, we will explain the genetic operators. The one point crossover [6] of two individuals is performed as follows:

Given two parents $P_1$ and $P_2$ as:

$$P_1 = [p_{11}, p_{12,...,}p_{1N}], \quad P_2 = [p_{21}, p_{22}, ..., p_{2N}],$$

we choose a crossover point at random, in such a way that the new individual (child1) will inherit all positions to the left of the crossover point from $P_1$, and the positions to the right from $P_2$. The following example illustrates this crossover operator:

$P_1 =$ |**23**|**309**|276|12|513|, $P_2 =$ |506|281|**105**|**33**|**447**| $child1 =$ |23|309|105|33|447|

In this example, the crossover point is chosen between the second and third position, so the first two positions are copied from the first parent and the other three from the second.

To mutate an individual, we chose one of the $N$ initial positions at random and replace it with another position generated also at random, in the range between 1 and $T - l + 1$. The resulting pseudocode for the procedures is given in Algorithm 1. Notice that generational replacement is used as the survival selection strategy.

**Algorithm 1.** PbGA

*Input:* Number of strings $(N)$, string length $(T)$, motif length $(l)$, number of mutated elements $(d)$, mutation rate $(P_m)$, crossover rate $(P_c)$, Population Size $(PopSize)$, tournament size $(J)$, the maximum number of iterations without changes in the fitness of the best individual $(Max\_Iter\_No\_Change)$.

*Output:* A set of positions $\{p_1, p_2, \cdots, p_N\}$ and its corresponding score.

1 Initialize $PopSize$ individuals with random initial positions

2 Evaluate the fitness of the $PopSize$ individuals and the best individual is taken as the super individual.

3 While the number of generations without changes in the fitness is less   than $Max\_Iter\_No\_Change$, Do:

4    For i=1 to $PopSize$

5       Choose $J$ individuals to participate in a match and the winner will be Parent1

6       Choose $J$ individuals to participate in a match and the winner will be Parent2.

7       Perform crossover between Parent1 and Parent2 with probability $P_c$ to obtain a new individual

8       Apply mutation to each new individual with probability $P_m$.

9    Replace the actual population with the new $PopSize$ individuals

10   Evaluate the fitness of the new population to choose the best individual and compare it with the old one, then keep the best.

11 End While

## 3.2   Motif Based Encoding Genetic Algorithm: MbGA

A clear limitation with the initial positions based representation is that it is not trivial to deal with repetitions of occurrences in one or more sequences, but specially to deal with the case where a given sequence does not contain an occurrence at all. To improve the performance of the position based encoding, we borrow some ideas from the Pattern Branching Algorithm (PBA) [15]. The main ideas with the PBA are: to work with the motif itself not with the positions, to compute the quality of solution as a measure of the Hamming distance between the candidate motif and the closest length $l$ substring in each sequence. We take in our MbGA these two ideas. To do this, we change the representation of the individual, in such a way that it represents the motif (as in [10] and GA2 [21]), not the position of each occurrence as in [5], nor each motif occurrence as in [12]. If we use the position based approach the search space has a size of $(T - l + 1)^N$. It is clear that this size increases exponentially with the number of sequences $N$. If we choose the motif based representation the search space size is still exponential, $4^l$, although independent of the number and length of sequences, it grows exponentially with the motif length. However, in the planted motif problems the number and length of sequences is a real issue, for details see [4]. Notice that the difference in size of the search spaces will probably require a bigger population size for the position based encoding.

In the following example we see an individual which represents a candidate motif of length $l = 10$:

$$\mathbf{i} = \boxed{a|a|g|t|t|c|a|c|c|g}$$

Notice that this representation has no limitations to deal with repeats or with the absence of occurrences in a given sequence as the PbGA does.

To calculate the fitness of an individual we use the concept of *total distance* [15]. In order to compute this distance we first recall some basic definitions. The Hamming distance $d(s_1, s_2)$ between two substrings of length $l$ is the number of characters in which they differ. For each string $S_j$, let $d(\mathbf{i}, S_j) = min\{d(\mathbf{i}, p)|p \in S_j\}$, where $p$ denotes a substring of length $l$ and $\mathbf{i}$ the candidate motif. Then the total distance from $\mathbf{i}$ to the $N$ strings is given by $d(\mathbf{i}, S) = \sum_{j=1}^{N} d(\mathbf{i}, S_j)$. In this case we need to minimize this distance, so at the end of the algorithm, those who have survived will represent a candidate motif for the one that was originally implanted. It is not hard to see that the cost for evaluating this objective function is $O(T * l * N)$. This evaluation cost is more expensive than the one in PbGA which is only $O(l * N)$.

The genetic operators for this algorithm are similar to the ones used in the standard PbGA. For the recombination operator, we choose a crossover point at random. For the mutation, we choose one of the $l$ characters at random and replace it with another one, also generated at random from the alphabet {a,c,g,t}, where each character is equally likely to be selected.

The resulting algorithm structure is similar to the one introduced in Algorithm 1, with the difference in the representation and the objective function for which we want to minimize the total distance instead of maximizing the total score. The output for this algorithm is a candidate motif of length $l$.

In order to compare this approach with the PbGA we have designed a set of computational experiments which are described in the next section.

## 4    Experimental Setup and Results

To test both algorithms, we generated $(l, d)$-planted motif instances of (10,2), (10,3), (11,2), (11,3), (12,3), (12,4), (15,4), (16,5), (18,6), (20,7), (30,11) and (40,15) as follows: first, a motif $M$ of length $l$ is generated by choosing $l$ bases at random. Second, $N = 20$ occurrences of the motif are created by randomly choosing $d$ positions per occurrence (without replacement) and replacing the base by a randomly chosen one. Third, we construct $N$ background sequences each of length $T = 600$ selecting the bases at random. Finally, we assign each occurrence to a random position in a background sequence, one occurrence per sequence.

After a non extensive trial and error tuning process, we have selected the algorithms' parameters as follows. For the PbGA we set the population size to $1,000$ individuals, selecting three individuals for each tournament ($J = 3$), and using a mutation probability of $P_m = 0.4$ and a crossover probability of $P_c = 1.0$. In the MbGA we set the population size also to $1,000$ individuals in an attempt to equalize the computation times, define a tournament size of $J = 3$, and set the mutation and crossover probabilities as in the PbGA. The parameter $Max\_Iter\_No\_Change$ equals 100 for the PbGA and 15 for the MbGA. Both algorithms are run 30 times for each instance. These algorithms were implemented in C++ and compiled with g++. We run both algorithms over all instances on a PC with AMD Athlon 3000+ processor with 512MB in RAM, with Linux kernel $2.6.12 - 9$amd64.

Since the MbGA generates a candidate motif and the PbGA outputs candidate positions for occurrences of the motif on each sequence $S_i$, we need to generate a candidate motif for the PbGA or a set of candidate positions for the MbGA. We decided to adopt the second option. The way to generate the candidate positions is just by taking the candidate motif generated by MbGA and selecting the closest (Hamming distance) length $l$ substring on each sequence $S_i$. Then the starting position of each of these substrings is given as output. Once this is done the corresponding score is computed.

Table 1 compares the average solution quality (total score) of both algorithms for all instances. The first column indicates the specific instance name, the second column gives the average objective function values for the PbGA, the third column the average objective function for the MbGA, and columns third and fourth their respective standard deviations. Table 2 compares both average computation times and their respective standard deviations. The values shown in parenthesis are the best in 30 runs, and the ones in bold are the best found for the respective instance, which are, at the same time, the optimal solution.

**Table 1.** Comparison of average total score ($\overline{F}$) and their corresponding standard deviation. PbGA stands for Position Based Genetic Algorithm and MbGA for Motif Based Genetic Algorithm. The score in parenthesis is the best score found by the algorithm in 30 runs, and the ones in bold are the score of the best solution found that corresponds to the optimum.

| $Instance(l,d)$ | $\overline{F}$PbGA | $\overline{F}$MbGA | S.D. PbGA | S.D. MbGA |
|---|---|---|---|---|
| (10,2) | 144.07(151) | 157.8(**166**) | 3.37 | 3.6 |
| (11,2) | 153.37(160) | 178.47(**193**) | 3.47 | 8.41 |
| (12,3) | 167.07(201) | 196.33(**210**) | 7.97 | 6.16 |
| (15,4) | 195.97(232) | 239.33(**250**) | 7.69 | 8.55 |
| (16,5) | 204.18(211) | 240.57(**254**) | 4.68 | 10.6 |

The results in Table 1 show that the motif based algorithm obtains most of the time, for the instance $(10, 2)$, solutions close to the optimum. This means that the motif found at each execution differs from the real motif in one or two bases only. The position based algorithm was not able to find the optimal solution in neither of the 30 executions. In the second instance we can see similar results as in the first one.

**Table 2.** Computation time average ($\overline{T}$) and its standard deviation for the PbGA and the MbGA algorithms

| $case(l, d)$ | $\overline{T}$PbGA | $\overline{T}$MbGA | S.D. PbGA | S.D. MbGA |
|---|---|---|---|---|
| (10,2) | 11.1 | 81.67 | 2.58 | 17.95 |
| (11,2) | 11.77 | 91.8 | 3.05 | 21.57 |
| (12,3) | 12.4 | 131.5 | 2.34 | 28.92 |
| (15,4) | 31.1 | 155.17 | 10.66 | 55.55 |
| (16,5) | 12.39 | 335.1 | 2.35 | 95.52 |

**Table 3.** Comparison of average total score ($\overline{F}$) for SGA allowed extra time, the PbGA, and the Gibbs sampler. The score in parenthesis is the best score found by the algorithms in 30 runs, and the ones in bold are the score of the best solution found.

| $instance(l, d)$ | $\overline{F}$PbGA | $\overline{F}$MbGA | Gibbs Sampler |
|---|---|---|---|
| (10,2) | 147.1(**166**) | 157.8(**166**) | 160 |
| (10,3) | 146.17(151) | 155.57(**158**) | 144 |
| (11,2) | 155.63(159) | 178.47(**193**) | **193** |
| (11,3) | 155.97(173) | 173.67(**176**) | **176** |
| (12,3) | 167.6(**210**) | 196.33(**210**) | **210** |
| (12,4) | 166(172) | 178.2(**184**) | **184** |
| (15,4) | 197.3(233) | 239.33(**250**) | **250** |
| (16,5) | 207.67(246) | 240.57(**254**) | **254** |
| (18,6) | 221.87(231) | 238.53(**279**) | 209 |
| (20,7) | 242.03(249) | 265.5(**300**) | 296 |
| (30,11) | 335.97(407) | 375.93(**454**) | 314 |
| (40,15) | 490.5(**613**) | 584.13(**613**) | **613** |

**Table 4.** Comparison of the performance achieved by MbGA on a real instance. The CRP binding-site [20].

| Type of Sol. | Initial Positions | Score |
|---|---|---|
| True Pos.: | 61,17 55,17 76 63 50 7,60 42 39 9,80 14 61 41 48 71 17 53 1,84 78 | 240 |
| PbGA Best: | 61 55 31 63 81 7 24 66 9 14 61 41 48 71 17 53 5 79 | 241 |
| PbGA Dev.: | 0, 0, -45, 0, 31, 0, -18, 27, 0, 0, 0, 0, 0, 0, 0, 0, 4, 1 | |
| MbGA Best | 45 55 76 63 50 7 24 66 9 14 29 41 48 71 6 53 75 8 | 246 |
| MbGA Dev.: | -16, 0, 0, 0, 0, 0, -18, 27, 0, 0, -32, 0, 0, 0, -11, 0, -9, -50 | |

In the instance $(12, 3)$ again the MbGA were able to find the optimal solution. In the instance $(15, 4)$, the MbGA obtained a very good performance, since all its solutions were close to the best value (as it happened with the instance $(10, 2)$). In contrast with what happens with MbGA, the standard algorithm has a lower performance, since the objective function average is far from the optimum. In the last instance $(16, 5)$, we can see that MbGA results surpass those obtained by the standard one. In all cases we can see that there is a clear improvement of the MbGA over the standard GA in terms of solution quality. If we consider the improvement on the average values we have a 9.5% improvement for the $(10, 2)$, 16.36% for the $(11, 2)$, 17.51% for the $(12, 3)$, 22.13% for the $(15, 4)$, and 17.82% for the $(16, 5)$. It is clear that as the motif length increases (with the exception of (16,5)) the outperformance of the MbGA over the PbGA also increases.

Another very important issue is the computation time. In Table 2 we can see that the computation time increases as the motif length increases in both algorithms. However, the position based algorithm takes less time in all instances. The reason of these results is that the objective function is computationally more expensive in the MbGA than in the PbGA.

Another set of experiments were included allowing the PbGA to have the same or more computation time than the MbGA by increasing the population size from 1000 to 3000 individuals in the PbGA. The results are shown in Table 3. We can see here that no significant improvement is achieved by the PbGA. In the same table results given by the Gibbs sampler [11] are also shown. We can see that the MbGA presents competitive results, outperforming the Gibbs sampler in five out of twelve instances. These results showing that MbGA is better than PbGA support the encoding selection made by other authors [10].

We also deal with a real problem consisting of 18 sequences of 104 base pairs each. There are repeats of the motif in 5 of the 18 sequences. This is known as the CRP binding-site and was taken from [20], and also used in [5]. Table 3 shows the results for this case. In the first row we have the actual starting positions of the motifs along with its score. In the second row we have the positions given by the highest score individual of the PbGA. In the third row we have the error of each predicted occurrence. That is, if the starting position for the occurrence predicted by the algorithms coincides with the actual position of that occurrence this error is zero. Otherwise, if the predicted position if shifted to the right then the error is positive, and when this predicted position is shifted to the left the error is negative. The same information is reported for the MbGA in rows four and five.

PbGA and MbGA produce similar results; PbGA predicts one more position than MbGA does. However, MbGA obtains the best result in terms of the score. It is interesting to note that the algorithms produce higher scores than the one produced by the real motifs. This fact motivates to search for an objective function that better captures the real motifs. This result is basically saying that the objective function does not strongly discriminate the signal (motif) from the background noise.

## 5   Conclusions and Future Work

We have proposed an experimental comparison of two encoding schemes for the implanted motif finding problem. A position based and a motif based encodings. The position based encoding takes as chromosome a set of initial positions for occurrences in each sequence; the motif based takes as a chromosome a candidate motif. The second encoding better captures the idea of and individual and its evolution regarding biological motifs. This algorithm is based on ideas provided by a successful heuristic for the planted $(l, d)$-motif model known as Pattern Branching. Experimental preliminary results show a clear solution quality improvement of the motif based representation over the position based representation.

Future immediate research will be aimed at extending the set of instances in the comparison set. We need also to apply the motif based algorithm to real data and compare its performance with respect to other algorithms, like Random Projections and Pattern Branching itself. We will also improve our implementation to decrease the computation time, and to search for a more suitable objective function.

# References

1. Bailey, T., Elkan, C.: Unsupervised learning of multiple motifs in biopolymers using expectation maximization. Machine Learning 21, 51–80 (1995)
2. Blanchette, M., Schwikowski, B., Tompa, M.: Algorithms for philogenetic footprinting. J. Comp. Biol. 9, 211–223 (2002)
3. Brazma, A., Jonassen, I., Vilo, J., Ukkonen, E.: Predicting gene regulatory elements *in silico* on a genomic scale. Genome Res. 15, 1202–1215 (1998)
4. Buhler, J., Martin, T.: Finding Motifs Using Random Projections. Journal of Computational Biology 9(2), 225–242 (2002)
5. Che, D., Song, Y., Rasheed, K.: MDGA: Motif Discovery Using A Genetic Algorithm. GECCO'05 (June 25-29, 2005)
6. Goldberg, D.E.: Genetic Algorithms in Search, Optimization and Machine Learning. Addison-Wesley, Reading (1989)
7. Gusfield, D.: Algorithms on Strings, Trees and Sequences: Computer Science and Computational Biolofy. Cambridge University Press, Cambridge (1997)
8. Hertz, G., Stormo, G.: Identifying DNA and protein patterns with statistically significant alignments of multiple sequences. Bioinformatics 15, 563–677 (1999)
9. Jones, N.C., Pevzner, P.A.: Introduction to Bioinformatics Algorithms. MIT Press, Cambridge (2004)
10. Karaoglu, N., Maurer-Stroh, S., Manderick, B.: GAMOT: An efficient genetic algorithm for finding challenging motifs in DNA sequences. In: Apostolico, A., Guerra, C., Istrail, S., Pevzner, P., Waterman, M. (eds.) RECOMB 2006. LNCS (LNBI), vol. 3909, Springer, Heidelberg (2006)
11. Lawrence, C., Altschul, S., Bogusky, M., Liu, J., Neuwald, A., Wootton, J.: Detecting subtle sequence signals: a Gibbs sampling strategy for multiple alignment. Science 208–214 (1993)
12. Liu, F.M., Tsai, J.P., Chen, R.M., Chen, S.N., Shih, S.H.: FMGA: finding motifs by genetic algorithm. In: IEEE Fourth Symposium on Bioinformatics and Bioengineering (BIBE 2004), pp. 459–466. IEEE Computer Society Press, Los Alamitos (2004)
13. Liu, X., Brutlag, D.L., Liu, J.S.: BioProspector: discovering conserved DNA motifs in upstream regulatory regions of co-expressed genes. Pac. Symp. Biocomput. 6, 127–138 (2001)
14. Pevzner, P., Sze, S.-H.: Combinatorial approaches to finding subtle signals in DNA sequences. In: Proc. 8th Int. Conf. Intelligent Systems for Molecular Biology, pp. 269–278 (2000)
15. Price, A., Ramabhadram, S., Pevzner, P.: Finding Subtle Motifs by Branching from Sample Strings. Bioinformatics 1(1), 1–7 (2003)

16. Roth, F.R., Hughes, J.D., Estep, P.E., Church, G.M., Finding, D.N.A.: Regulatory Motifs within unaligned non-coding sequences clustered by whole-Genome mRNA quantitation. Nature Biotechnology 16(10), 939–945 (1998)
17. Sagot, M.-F.: Spelling approximate repeated or common motifs using a suffix tree. In: Lucchesi, C.L., Moura, A.V. (eds.) LATIN 1998. LNCS, vol. 1380, pp. 111–127. Springer, Heidelberg (1998)
18. Sagot, M.-F., Escalier, V., Viari, A., Soldano, H.: Searching for repeated words in a text allowing for mismatches and gaps. In: Baeza-Yates, R., Manber, U. (eds.) Second South American Workshop on String Processing, Viñas del Mar, Chili, pp. 87–100. University of Chili (1995)
19. Sinha, S., Tompa, M.: A statistical Method for finding transcription factor binding sites. In: Proc. 8th Int. Conf. Intelligent Systems for Molecular Biology, pp. 344–354 (2000)
20. Stormo, G.D., Hartzell III, G.W: Identifying protein-binding sites from unaligned DNA fragments. PNAS 86, 1183–1187 (1989)
21. Stavrovskaya, E.D., Mironov, A.A.: Two genetic algorithms for identification of regulatory signals. In: Silico Biology (2003)
22. Waterman, M.S., Arratia, R., Galas, D.J.: Pattern recognition in several sequences: consensus and alignment. Bull. Math. Biol. 46, 515–527 (1984)
23. Zaslavsky, E., Singh, M.: A combinatorial optimization approach for diverse motif finding applications. Algorithms for Molecular Biology 1–13 (2006)

# Multi-Objective Clustering Ensemble with Prior Knowledge*

Katti Faceli[1], André C. P. L. F. de Carvalho[1], and Marcílio C. P. de Souto[2]

[1] Universidade de São Paulo
Instituto de Ciências Matemáticas e de Computação
Departamento de Ciências de Computação e Estatística
Caixa Postal 668, 13560-970 – São Carlos, SP, Brasil
{katti,andre}@icmc.usp.br
[2] Universidade Federal do Rio Grande do Norte
Departamento de Informática e Matemática Aplicada – DIMAp
Campus Universitario, 59072-970 – Natal, RN, Brazil
marcilio@dimap.ufrn.br

**Abstract.** In this paper, we introduce an approach to integrate prior knowledge in cluster analysis, which is different from the existing ones for semi-supervised clustering methods. In order to aid the discovery of alternative structures present in the data, we consider the knowledge of some existing complete classification of such data. The approach proposed is based on our Multi-Objective Clustering Ensemble algorithm (MOCLE). This algorithm generates a concise and stable set of partitions, which represents different trade-offs between several measures of partition quality. The prior knowledge is automatically integrated in MOCLE by embedding it into one of the objective functions. In this case, the function gives as output the quality of a partition, considering the prior knowledge of one of the known structures of the data.

## 1 Introduction

Cluster analysis has been largely employed to address different kinds of problems in Bioinformatics, ranging from the identification of genes function to the discovery of groups and subgroups of diseases [1,2]. As an unsupervised learning task, such an analysis does not take into account a previously known classification of the data (prior knowledge): it relies only on the similarities of the objects. However, in the context of discovering disease subtypes, for instance, in order to establish if clusters generated correspond to actual disease subtypes, the domain experts compare them to a known classification of the data [3,4,5].

In fact, cluster analysis present two main difficulties: one regards the use of a priori knowledge to verify the findings, and the other is related with the inherent complications related with cluster analysis. The automatic use of prior knowledge has been addressed by semi-supervised clustering techniques [6,7].

---

* This work was supported by FAPESP and CNPq.

M.-F. Sagot and M.E.M.T. Walter (Eds.): BSB 2007, LNBI 4643, pp. 34–45, 2007.

However, these techniques consider that only a small number of the objects in the data are labeled. They try to achieve a performance higher than those achieved by either pure supervised or unsupervised techniques in the discovery of one structure partially known.

An inherent complication in cluster analysis is the lack of a precise definition for what a cluster is [8]. This results in a large number of clustering algorithms, each one looking for clusters according to a different cluster definition (or clustering criterion) [9]. Moreover, clustering algorithms can find structures (partitions) at different refinement levels (different numbers of clusters or cluster densities), depending on their parameter settings [10].

Thus, one of the main difficulties of cluster analysis is the selection of the best model for a given dataset. Clustering validation techniques support this task. However, most of them are biased towards a clustering criterion [11]. Also, each algorithm looks for a homogeneous structure (all clusters conforming to the same cluster definition), while data can present an heterogeneous structure (each cluster conforming to a different cluster definition) [9].

Another issue is that the same data can have more than one relevant structure, each one representing a different interpretation of the data [12]. The usual application of cluster analysis to explore a dataset focuses on the discovery of only one structure that best fits the data. This limits the amount of knowledge that could be obtained with cluster analysis.

Cluster ensemble and multi-objective clustering approaches have been employed to address all the difficulties previously described [13,12]. More recently, in [14], we proposed an integration of the ideas of these approaches in our Multi-objective Clustering Ensemble (MOCLE) [14]. The idea is not only minimize the intrinsic problems of cluster analysis, but also the limitations of the cluster ensemble and multi-objective clustering methods when used separately.

The essence of MOCLE is the simultaneous optimization of different clustering validation measures (objective functions) using a Pareto-based multi-objective genetic algorithm together with a special crossover operator. For instance, the prior knowledge about a known structure of the data can be integrated into MOCLE by means of an additional objective function that takes external information into account. The result of MOCLE is a concise, stable and high quality set of partitions representing different trade-offs between the validation measures optimized.

Techniques like MOCLE are very useful in functional genomics and gene expression data analysis, where the data usually have multiple meaningful interpretations: genes can fit into more than one functional category or a disease, like cancer, can present, depending on the required level of investigation, different subtypes [5,4,3]. Furthermore, robustness against different data conformation is a key issue in these areas, as there is not much previous knowledge to guide the choice of the algorithms or the parameters configurations and the structures present in the data tend to be complex.

In this paper, we discuss how prior knowledge of one complete classification of the data can be automatically used to aid in the discovery of other structures.

This task, as previously mentioned, is accomplished with the optimization of an extra objective function by MOCLE. This aim is different from that of classical semi-supervised algorithms, which try to obtain the most precise model (in terms of classification error) when compared to those models generated by pure supervised and unsupervised techniques.

## 2   Related Work

The semi-supervised clustering techniques make use of prior knowledge on the problem domain to guide the clustering process [7,6]. They are often suitable for data with very limited previous knowledge, that is, for which a very small subset of the objects are labeled, while the class of most of the objects are unknown.

Classes are entities related to categories previously defined in the real world to organize the objects. Clusters, on the other hand, are entities defined by the application of mathematical/statistical concepts to the data. The classes can be related to one or more of the mathematical/statistical concepts, but this need not to be the case. They do not necessarily correspond to the clusters. The semi-supervised clustering techniques assume that the classes and clusters are consistent, complement each other and their combined use can improve the classification accuracy [7]. If these assumptions are strongly violated, the improvement cannot be guaranteed.

There are several ways to integrate prior knowledge into the clustering process. For instance, the closet approaches to ours do this by integrating the knowledge through the adaptation of the clustering criterion (or objective function) [7,6]. Demiriz et al. [6], for example, proposed a method to cluster the data whose goal is to generate the purest possible clusters regarding the class distribution. They use a genetic algorithm to minimize a function that is a linear combination of a measure of dispersion of the clusters (unsupervised) and a measure of impurity with respect to the known classes (supervised). As dispersion measure, they investigated the within cluster variance and the Davies-Bouldin index [10]. The Gini index was used as the impurity measure [15].

Another interesting work in this area is the one by Handl and Knowles [7]. They extended their algorithm MOCK (Multi-Objective Clustering with automatic K-determination) to consider prior knowledge. Originally, MOCK simultaneously optimize two complementary objectives: overall deviation and connectivity. In [7], the authors employ an additional objective function that takes external knowledge into account: the corrected Rand index (CR) [10]. The CR is computed using only the labeled data. Handl and Knowles also present an alternative with two objectives: the silhouette width as the unsupervised objective and the CR as the supervised one.

As previously mentioned, these techniques and ours have distinct purposes. The previous techniques aim at the improvement of the accuracy of a model generated to classify new objects, if compared to the models that would be generated with the purely supervised or unsupervised techniques. In contrast, our focus is on exploratory data analysis, where we search for several useful descriptions of the data. In fact, our main aim in this paper is to analyze if the use of external knowledge can improve the capacity of MOCLE.

# 3  Multi-Objective Clustering Ensemble - MOCLE

In order to try to overcome the difficulties of the traditional algorithms for cluster analysis, MOCLE combines characteristics from both the cluster ensemble and multi-objective clustering methods [14]. As any cluster ensemble, MOCLE is composed of two main steps: (1) generation of a diverse set of base partitions and (2) determination of the consensus partition. Our approach differs from cluster ensemble methods in two ways. First, we look for a set of "consensus" partitions instead of only one. In fact, our set of solutions may contain partitions that are combinations of other partitions, or partitions of high quality that already appeared in the set of individual partitions. Second, we combine pairs of partitions, iteratively, in an optimization process, instead of the usual combination of all partitions at the same time. Such an iterative combination/selection of the partitions avoids the negative influence of low quality base partitions that can decrease the quality of the results of the traditional ensembles.

More precisely, MOCLE works as follows. Initially, a set of base partitions is generated. Conceptually different clustering algorithms, optimizing different clustering criteria, are employed for this purpose. For example, algorithms that look for compact clusters should be used together with algorithms that look for connected clusters. The more diverse the algorithms are, the larger the number of types of cluster that can be discovered. Several parameter settings for the algorithms are also considered in the construction of the set of base partitions. This generates partitions with clusters at different refinement levels (partitions with different numbers of clusters or partitions with clusters of several densities, for example). It is important to have partitions with different types of clusters at several refinement levels so that MOCLE can receive as much information as possible to find the largest number of possible existing structures. In fact, we assume that the relevant structures will be among the base partitions.

After generating the base partitions, the set of "consensus" partitions are found by the optimization of different objective functions using a Pareto-based multi-objective genetic algorithm. Any known algorithm can be employed. In this paper, we used the algorithm NSGA-II (Non-dominated Sorting Genetic Algorithm) [16]. The use of this class of genetic algorithm results in, as previously mentioned, a set of partitions, instead of a single partition produced by traditional and cluster ensemble methods. This is an important feature in domains like Bioinformatics, where the same data can have several interpretations.

The base partitions constitute the initial population to be used with the genetic algorithm. Each partition is an individual and is represented by an array of sets. Each set, in its turns, represents a cluster and contains the labels of its objects. In addition to the special initial population, two other adaptations are made in the traditional genetic algorithm: a special crossover operator and the use of diverse clustering validation measures as objective functions. Together with the initial population, our special crossover operator is responsible for the ensemble aspect of MOCLE. This operator finds the consensus between two parent partitions. Any existing cluster ensemble method that can be applied to a pair of partitions can be used as our crossover operator.

In this paper, we use the Hybrid Bipartite Graph Formulation (HBGF) as the crossover [17]. HBGF is based on graph partitioning. In this method, first, a bipartite graph is constructed using the set of base partitions, modeling their objects and clusters simultaneously as vertices. Next, the graph is partitioned by a traditional graph partitioning technique. The resulting division of the objects is the consensus partition. To use HBGF as our crossover operator, we select two parents by binary tournament. The number of clusters of the resulting consensus partition is randomly chosen in the interval of variation of the numbers of clusters of the parents. Next, we apply HBGF to combine the parent partitions, generating a consensus partition with the number of clusters chosen.

With this operator, the partitions are combined in pairs, iteratively, during the evolution process. The consensus partitions, generated at each iteration, are also considered in the next combinations. This iterative combination avoids the negative influence of the low quality partitions present in most of the traditional cluster ensemble methods. The low quality partitions are gradually eliminated, while the best individual partitions and the good combinations are maintained for further combination.

Since we want to restrict the search space to the base partitions and their combination, we do not apply a mutation operator. Therefore, the genetic algorithm aims to select the best partitions, and not to explore all the space of possible partitions. In the pure Pareto based multi-objective clustering scenario, differences in the assignment of only one object to a different cluster in two partitions can result in a different trade-off of the measures optimized. This can result in a high number of very similar partitions in the approximation of the Pareto front obtained.

In contrast, we argue that, in the context of clustering, the aim should not be the generation of the most complete Pareto front approximation possible. Indeed, having solutions representing each region of the Pareto front is enough to provide a relevant set of alternative partitions. Considering this fact, MOCLE aims at the generation of a concise set of solutions that are representative of the Pareto front. As already mentioned, MOCLE relies on the ability of the clustering algorithms in finding high quality partitions according to the employed criteria. Starting with a set of potentially good partitions, MOCLE uses the multiple objectives to select the best compromises. New partitions are created only by means of the crossover operator and represent the consensus among other existing partitions. As our crossover operator only produces combinations of existing partitions and no mutation is used, the search space will not be explored in details. Thus, the large amount of similar partitions will not be produced by MOCLE, favoring the concision of the set of solutions obtained.

Finally, the objective functions should measure the quality of partitions in different ways, each one related to a different clustering criterion. They should also complement each other. For the completely unsupervised case, we have used the same measures employed in [12]: overall deviation and connectivity. The overall deviation of a partition measures the overall summed distances between objects and their corresponding cluster center. This measure is strongly biased towards spherically shaped clusters and improves with the increase in the number of clusters. The connectivity reflects how often neighboring objects have been placed in the same cluster. It improves with the decrease in the number of clusters. The connectivity is able to detect arbitrarily shaped clusters, but it is not robust to deal with overlapping clusters. These two objectives, to be minimized, balance each other's tendency to increase or decrease the number of clusters, avoiding the convergence to trivial solutions. The objective functions are responsible for the selection of the high quality partitions and the robustness of MOCLE with respect to different data conformation.

The prior knowledge about a known structure of the data can be integrated into MOCLE by means of an additional objective function that takes external information into account. In this paper we investigated the information gain measure for this purpose [15,18]. The CR index, used in [7], would not be appropriate for our purpose. It considers negatively the subdivisions of clusters, while we want to find partitions that are refinement of the known partition. As in Demiriz et al. [6], we aim at generating partitions with clusters as pure as possible regarding the class distribution. When used as criteria to the division of nodes in decision trees, the Gini index [6] and the information gain produce very similar trees [18]. However, the Gini index gives preference to divisions that place the largest class in one pure node and all other classes in another node [15]. Information gain, in its turn, favors the generation of nodes with balanced sizes. The application of the Gini index to our problem would favor partitions with a large pure cluster (considering the known structure) and a cluster that mixed the other classes, in detriment of partitions with subdivisions of the large class together with a good separation of the smaller classes, that would be preferred. Therefore, we decided to use the information gain.

## 4    Experiments

In order to evaluate our approach, we choose datasets that contain more than one possible structure. $\Pi_E = \{\pi^{E1}, \pi^{E2}, ..., \pi^{En^E}\}$ is the set of known structures for a given dataset, where $n^E$ is the number of known structures and $\pi^{Ej}$ is the $j$th known structure. Three artificial and two real datasets were employed. Table 1 summarizes the main characteristics of the datasets. In this table, $n$ is the number of objects, $d$ is the dimension of the dataset (number of attributes), $n^E$ is the number of known structures and $K^{Ej}$ is the number of clusters of the $j$th structure.

The artificial datasets can be seen in Fig. 1. They were designed to contain at least two distinguishing structures. These structures are heterogeneous and are in different refinement levels.

**Table 1.** Datasets characteristics

| Dataset | $n$ | $d$ | $n^E$ | $K^{E1}$ | $K^{E2}$ | $K^{E3}$ | $K^{E4}$ |
|---------|-----|-----|-------|----------|----------|----------|----------|
| ds2c2sc13 | 588 | 2 | 3 | 2 | 5 | 13 | - |
| ds3c3sc6 | 905 | 2 | 2 | 3 | 6 | - | - |
| ds4c2sc8 | 485 | 2 | 2 | 2 | 8 | - | - |
| golub | 72 | 3571 | 4 | 2 | 3 | 4 | 2 |
| leukemia | 327 | 271 | 2 | 3 | 7 | - | - |

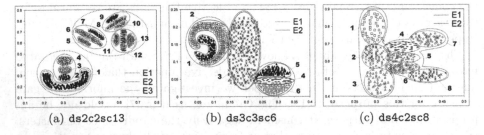

(a) ds2c2sc13          (b) ds3c3sc6          (c) ds4c2sc8

**Fig. 1.** Artificial datasets

For the real datasets, the different structures correspond to different known classifications of the data. Thus, we assume that the known classifications are in accordance with some of the clustering criteria we use. However, a classification could be unrelated to a clustering criterion. This would lead to a low performance for all clustering techniques.

The golub dataset contains gene expression data from acute leukemia patients [4]. For our analysis, we consider four distinct known structures of this dataset. The two main structures refer to types and subtypes of acute leukemia: E1 classifies the samples in Acute Lymphoblastic Leukemia (ALL) and Acute Myeloid Leukemia (AML). E2 contains a refinement of the ALL class. In this case, the data are classified in AML, T-ALL (T-lineage ALL) and B-ALL (B-lineage ALL). The other structures correspond to different types of information. E3 classifies the samples according to the institution where the samples came from: DFCI (Dana-Farber Cancer Institute), CALGB (Cancer and Leukemia Group B), SJCRH (St. Jude Children's Research Hospital) and CCG (Children's Cancer Group). E4 shows if the samples are from bone marrow (BM) or peripheral blood (PB). The data were preprocessed in the same way as in [4]. First, a floor of 100 and a ceiling of 16000 were applied. Then, we eliminated the genes with $max/min \leq 5$ and $(max - min) \leq 500$, where $max$ and $min$ refer respectively to the maximum and minimum expression levels of a particular gene across mRNA samples. Finally, a base 10 logarithmic transformation was applied.

The leukemia dataset, described in the literature as St. Jude leukemia, contains gene expression data related to subtypes of pediatric acute lymphoblastic leukemia [3]. For this dataset, we considered two structures. E1 classifies the objects in B-ALL, T-ALL and OTHERS (objects that does not fit into the other group). E2 contains a refinement of the class B-ALL and divides the objects in BCR-ABL, E2A-PBX1, "Hyperdiploid>50", MLL, TEL-AML1, T-ALL and OTHERS. We employed the pre-processed version of the dataset available at http://sdmc.lit.org.sg/GEDatasets/. Moreover, we selected the genes that best define each group (40 per group), identified by Yeoh et al. with the chi-square metric [3]. We also converted the attributes to the interval $[0, 1]$. All this were made to use the data in the same way as in its original paper.

For all datasets, we generate the initial population with the algorithms k-means (KM), average-link (AL), single-link (SL) [10] and Shared Nearest Neighbors (SNN) [19]. These algorithms generate different types of clusters. KM and LM looks for compact clusters and SL and SNN obtain connected clusters. KM, LM and LS were chosen because they are traditional and largely employed clustering algorithms [10]. In its turns, SNN is a recent technique and was selected because it can robustly deal with high dimensionality, noise and outliers [19]. In order to consider different refinement levels, we adjust the parameters to generate partitions with numbers of clusters, $k \in [K^{min}, K^{max}]$, where $K^{min} = \min_{\pi^{Ej} \in \Pi_E} K^{Ej}$ (the smallest number of clusters among those of the known structures) and $K^{max} = 2 \max_{\pi^{Ej} \in \Pi_E} K^{Ej}$. This procedure generates an initial population of different size for each dataset.

To minimize the occurrence of suboptimal solutions, we run KM 30 times for each value of $k$, each time with a random choice of initial centers. The partition with the lowest squared error for each value of $k$ was chosen to take part of initial population. The AL and SL partitions were obtained by generating the trees and cutting them in order to produce one partition for each value of $k$. For SNN, we run the algorithm with several values for the parameters $NN$ (2%, 5%, 10%, 20%, 30% and 40% of $n$), $topic$ (0, 0.2, 0.4, 0.6, 0.8 and 1) and $merge$ (0, 0.2, 0.4, 0.6, 0.8 and 1). In preliminary experiments, we noticed that varying the other parameters did not produce very different results. Thus, the default value was used for the parameter $strong$, and the value 0 was used for the parameters $noise$ and $label$ (to have all points assigned to a cluster and no point excluded as noise). From the partitions created with these parameters values, we chose only those partitions having $k$ in the interval of interest to be used in the initial population.

In the experiments, we compare two versions of MOCLE: one unsupervised (MUH) and one that take prior knowledge into account (MSH). The only difference between them is that MSH includes the third objective function that depends on previous knowledge. The same initial configuration, including the initial population, was used for both versions. As MOCLE is not deterministic, it was run 30 times with the same initial configuration. For all datasets, the known structure presented to MSH as prior knowledge was E1. The number of generations used was set to 50. Preliminary experiments showed that increasing the number of generations did not modify the Pareto front approximation obtained. The internal population size, $n^I$, depends on the number of partitions

**Table 2.** Parameter values

| Dataset | $K^{min}$ | $K^{max}$ | $v$ | $NN$ | $n^I$ |
|---------|-----------|-----------|-----|------|-------|
| ds2c2sc13 | 2 | 26 | 30 | 12, 29, 59, 176, 235 | 114 |
| ds3c3sc6 | 3 | 12 | 46 | 18, 45, 91, 272, 362 | 42 |
| ds4c2sc8 | 2 | 16 | 25 | 10, 24, 49, 146, 194 | 66 |
| golub | 2 | 8 | 4 | 1, 4, 7, 22, 29 | 46 |
| leukemia | 3 | 14 | 17 | 7, 16, 33, 98, 131 | 38 |

generated by the individual algorithms having $k \in [K^{min}, K^{max}]$. The number of nearest neighbors used to calculate the connectivity, $v$, was set to 5% of $n$ (size of the dataset). Table 2 shows, for each dataset, the values of all parameters.

Besides comparing MUH and MSH, we also include the results of the individual algorithms (KM, AL, SL and SNN) as a reference. Also, as a reference, we run two other techniques that aims at the combination of different clustering criterion that are purely unsupervised: the multi-objective clustering of Handl and Knowles (MOCK) [12] and the ensemble (ES) of Strehl and Ghosh [13], with the equivalent parameters set to the same values used for MOCLE. The results of MUH, MSH and MOCK are a set of $n^S$ solutions, $\Pi_S = \{\pi^{S1}, \pi^{S2}, ..., \pi^{Sn^S}\}$, which is an approximation of the Pareto front. The individual algorithms do not generate a set of solutions. In order to compare their results with those of MOCLE, we considered the solutions generated for the initial population with a given algorithm as its set of solutions. That is, the partitions of the initial population that were generated with AL compose the set of solutions of the algorithm AL, for example. For KM, the results of each of the 30 runs count as a set of solutions. ES also do not generate a set of solutions. Thus, we construct this set by generating one solution for each value of $k \in [K^{min}, K^{max}]$.

## 5  Analysis of the Results

The results were evaluated via the Corrected Rand index ($CR$) [10]. This index measures the similarity between two partitions. Values of $CR$ close to 0 mean random partitions and close to 1 indicate a perfect match between the partitions. Table 3 shows the $CR$ or the average of $CRs$ for each technique, depending if the technique is deterministic or not. The highest values of $CR$ are highlighted in boldface.

For the deterministic techniques (AL, SL and SNN) we have just one set of solutions $\Pi_S$. In these cases, we calculated the $CR$ between each solution partition, $\pi^{Si} \in \Pi_S$, and each known structure, $\pi^{Ej} \in \Pi_E$. Next, for each known structure, $\pi^{Ej}$, we selected the best partition in $\Pi_S$ (the partition $\pi^{Si}$ with the highest $CR$ when compared with $\pi^{Ej}$).

For the non-deterministic techniques (KM, MOCK, ES, MUH and MSH), our experiments produced 30 sets of solutions, $\Pi_{S_1} ... \Pi_{S_{30}}$. For each set, $\Pi_{S_l}$, we calculated the $CR$ between each solution partition, $\pi^{S_l i} \in \Pi_{S_l}$, and each known structure, $\pi^{Ej} \in \Pi_E$. Next, for each known structure, $\pi^{Ej}$, we selected the best partition in each $\Pi_{S_l}$ (the partition $\pi^{S_l i}$ with the highest $CR$ when compared with $\pi^{Ej}$). Finally, for each known structure, we calculated the mean of the $CR$ of the 30 partitions selected (one for each $\Pi_{S_l}$).

**Table 3.** Performance of the algorithms

| Dataset | Structure | KM | AL | SL | SNN | MOCK | ES | MUH | MSH |
|---------|-----------|-----|-----|-----|-----|------|-----|------|------|
| ds2c2sc13 | E1 $(k = 2)$ | 1 | 1 | 1 | 1 | 1 | 0.691 | 1 | 1 |
|  | E2 $(k = 5)$ | 0.789 | 1 | 1 | 1 | 1 | 0.992 | 1 | 1 |
|  | E3 $(k = 13)$ | 0.651 | 0.617 | 0.872 | 1 | 0.708 | 0.786 | 0.777 | 0.777 |
| ds3c3sc6 | E1 $(k = 3)$ | 0.900 | 0.789 | 0.525 | 0.927 | 0.861 | 0.890 | 0.941 | **0.962** |
|  | E2 $(k = 6)$ | 0.601 | 0.490 | 0.250 | 0.618 | 0.648 | 0.597 | 0.674 | **0.697** |
| ds4c2sc8 | E1 $(k = 2)$ | **0.794** | 0.281 | 0.097 | 0.161 | 0.236 | 0.296 | 0.3486 | 0.3485 |
|  | E2 $(k = 8)$ | 0.816 | 0.831 | 0.019 | 0.586 | **0.885** | 0.846 | 0.856 | 0.861 |
| golub | E1 $(k = 2)$ | 0.507 | 0.876 | 0.078 | 0.855 | 0.684 | 0.743 | 0.876 | **0.900** |
|  | E2 $(k = 3)$ | 0.502 | 0.798 | 0.003 | 0.855 | 0.795 | 0.637 | 0.855 | **0.877** |
|  | E3 $(k = 4)$ | 0.402 | 0.693 | 0.108 | 0.677 | 0.622 | 0.589 | 0.693 | **0.699** |
|  | E4 $(k = 2)$ | 0.024 | 0.057 | **0.315** | 0.112 | 0.057 | 0.110 | **0.315** | 0.315 |
| leukemia | E1 $(k = 3)$ | **0.677** | 0.325 | 0.020 | 0.044 | 0.413 | 0.315 | 0.276 | 0.335 |
|  | E2 $(k = 7)$ | 0.748 | 0.543 | 0.005 | 0.004 | **0.782** | 0.659 | 0.770 | 0.775 |

In order to verify if the differences between the $CR$ of both versions of MOCLE were statistically significant, we performed the Wilcoxon signed-ranks test [20]. We found that the difference between MSH and MUH was statistically significant at a significance level of 0.05.

Looking at the individual algorithms, the first aspect that can be observed in Table 3 is that, for each dataset/known structure, a different individual algorithm showed the best performance. This illustrates the previously discussed difficulty in the choice of an appropriate algorithm to be used with a particular dataset.

We can also observe that both versions of MOCLE, in most of the cases (76.92%), obtained similar or better results than the best individual algorithm. There were only three cases where one of the individual algorithms performed better than MOCLE. This happened because the good solutions for these cases do not correspond to the best trade-offs of the objective functions. Considering that the structures investigated are in different refinement levels and are in accordance with different clustering criteria, the overall good performance of MOCLE shows its robustness against different data conformation.

Comparing both versions of MOCLE with either MOCK or ES, we observed that MOCLE obtained similar or better results than MOCK in 76.92% of the cases and outperforms ES in 92.31% of the cases.

We should observe that, for the structure E4 of the dataset `golub`, all techniques showed a very poor performance. This is clearly the case where the classification is not consistent with at least one of the clustering criteria optimized.

Finally, comparing MUH and MSH, we can see that in most of the cases, MSH performed similarly or better than MUH (92.31%). Among the 13 datasets/known structures we obtained four ties and just one case where MUH performed better. These 13 datasets/known structures include the structure E1 of each dataset (that is, the previous knowledge used in MSH). However, our main issue is to verify if the use of this knowledge helps in the obtaining of other structures, not considered as previous knowledge. In fact, we observed that, excluding the structure E1 of each dataset from the analysis, MSH performed similarly or better than MUH in 100% of the cases, from which 3 cases were ties. This shows the effectiveness of the integration of previous knowledge in MOCLE to help in the obtaining of other unknown structures.

## 6   Concluding Remarks

In this paper, we introduce an approach to integrate prior knowledge in cluster analysis, which is different from the existing ones for semi-supervised clustering methods. Our aim is to help the discovery of alternative structures that can be present in the data. The knowledge of some existing complete classification of such data is used for this purpose.

Our experimental results show that MOCLE, either considering previous knowledge or not, frequently selects the best results among those of the individual algorithms in a range of different data conformation/refinement levels. They also show that the use of prior knowledge with MOCLE (MSH) can result in solutions with higher quality than those obtained without it (MUH).

The use of other objective functions, unsupervised or not, could lead to further improvement of the results. Future works include the investigation of other functions. Another interesting direction for future research is the adaptation of MSH to the traditional semi-supervised clustering and its comparison with other existing techniques.

## References

1. Narayanan, E.K.A.: AIntelligent Bioinformatics: The Application of Artificial Intelligence Techniques to Bioinformatics Problems. John Wiley & Sons, Chichester (2005)
2. Wang, J.T.L., Zaki, M.J., Toivonen, H.T.T., Shasha, D.E. (eds.): Data Mining in Bioinformatics. Advanced Information and Knowledge Processing. Springer, Heidelberg (2003)
3. Yeoh, E.J., et al.: Classification, subtype discovery, and prediction of outcome in pediatric acute lymphoblastic leukemia by gene expression profiling. Cancer Cell 1(2), 133–143 (2002)
4. Golub, T., et al.: Molecular classification of cancer: Class discovery and class prediction by gene expression monitoring. Science 286(5439), 531–537 (1999)

5. Alizadeh, A., et al.: Distinct types of diffuse large B-cell lymphoma identified by gene expression profiling. Nature 403(6769), 503–511 (2000)
6. Demiriz, A., Bennett, K.P., Embrechts, M.J.: Semi-supervised clustering using genetic algorithms. In: Artificial Neural Networks in Engineering (ANNIE'1999), pp. 809–814 (1999)
7. Handl, J., Knowles, J.: On semi-supervised clustering via multiobjective optimization. In: Proceedings of the 8th annual conference on Genetic and evolutionary computation (GECCO'2006), pp. 1465–1472. ACM Press, New York, NY, USA (2006)
8. Xu, R., Wunsch, D.: Survey of clustering algorithms. IEEE Transactions on Neural Networks 16(3), 645–678 (2005)
9. Law, M., Topchy, A., Jain, A.K.: Multiobjective data clustering. In: Proceedings of the IEEE Computer Society Conference on Computer Vision and Pattern Recognition, vol. 2, pp. 424–430. IEEE Computer Society Press, Los Alamitos (2004)
10. Jain, A., Dubes, R.: Algorithms for Clustering Data. Prentice-Hall, Englewood Cliffs (1988)
11. Handl, J., Knowles, J., Kell, D.: Computational cluster validation in post-genomic data analysis. Bioinformatics 21(15), 3201–3212 (2005)
12. Handl, J., Knowles, J.: An evolutionary approach to multiobjective clustering. IEEE Transactions on Evolutionary Computation 11(1), 56–76 (2007)
13. Strehl, A., Ghosh, J.: Cluster ensembles - a knowledge reuse framework for combining multiple partitions. Journal on Machine Learning Research 3, 583–617 (2002)
14. Faceli, K., Carvalho, A., Souto, M.: Multi-objective clustering ensemble. In: Proceedings of the 6th International Conference on Hybrid Intelligent Systems (HIS'2006), Auckland, New Zealand, p. 51. IEEE Computer Society Press, Los Alamitos (2006)
15. Breiman, L.: Technical note: some properties of splitting criteria. Machine Learning 24(1), 41–47 (1996)
16. Deb, K., Pratap, A., Agarwal, S., Meyrivan, T.: A fast and elitist multi-objective genetic algorithm: NSGA-II. IEEE Transactions on Evolutionary Computation 6(2), 182–197 (2002)
17. Fern, X.Z., Brodley, C.E.: Solving cluster ensemble problems by bipartite graph partitioning. In: Proceedings of the Twenty-first International Conference on Machine Learning (ICML'2004), p. 36. ACM Press, New York, NY, USA (2004)
18. Raileanu, L.E., Stoffel, K.: Theoretical comparison between the Gini index and information gain criteria. Annals of Mathematics and Artiticial Intelligence 1(41), 77–93 (2004)
19. Ertöz, L., Steinbach, M., Kumar, V.: A new shared nearest neighbor clustering algorithm and its applications. In: Proceedings of the Workshop on Clustering High Dimensional Data and its Applications. 2nd SIAM International Conference on Data Mining (SDM'2002), pp. 105–115 (2002)
20. Demšar, J.: Statistical comparisons of classifiers over multiple data sets. JMLR 7, 1–30 (2006)

# Biological Sequence Comparison Application in Heterogeneous Environments with Dynamic Programming Algorithms

Marcelo N.P. Santana and Alba Cristina M.A. Melo

Department of Computer Science, Campus Universitario - Asa Norte, Caixa Postal 4466,
University of Brasilia, Brasilia – DF, CEP 70910-900, Brazil
{marcelo,albamm}@cic.unb.br

**Abstract.** This paper presents the design and evaluation of a task allocation framework for Biological Sequence Comparison applications that use dynamic programming and run in heterogeneous environments. The framework is composed by four modules and either task allocation policies or applications can be integrated to it. The results obtained with four different task allocation policies in a 10-machine heterogeneous environment show that, for some sequence sizes, we were able to reduce the execution time of the parallel application in 54.2%, with the appropriate allocation policy.

## 1 Introduction

Parallel processing has been extensively used to accelerate the production of results in many research areas such as meteorology and computational biology. However, the performance gains obtained by parallel computing are highly dependent on efficient task allocation mechanisms. Having a parallel application composed by a set of tasks and a set of processors, a task allocation algorithm will assign parallel tasks to processors observing some optimization criteria [7].

In its general formulation, the allocation problem is NP-complete [10]. For this reason, heuristic methods are generally used to find a good solution on a reasonable time. One traditional way to map tasks to processors is to apply an heuristic that makes decisions without taking into consideration the characteristics of the parallel application being scheduled. On the other hand, application-specific task allocators are targeted to a specific class of application and explore the applications characteristics to obtain more realistic mappings.

Nowadays, most of the parallel/distributed systems present a certain degree of heterogeneity. In this scenario, the resource allocation problem becomes more complex since the characteristics of the machines and networks that compose the parallel environment must be also taken into consideration. In the literature, there are examples of traditional task allocators/schedulers[15][13] and application-specific allocators [8][5] for heterogeneous environments.

In the last decade, genome projects have produced a very huge amount of biological data. In order to better understand a newly sequenced organism, biologists compare its sequence against millions of other organisms contained in genomic databases,

M.-F. Sagot and M.E.M.T. Walter (Eds.): BSB 2007, LNBI 4643, pp. 46–56, 2007.

in order to infer functional and structural properties. Sequence comparison is, thus, one of the most important mechanisms in computational biology.

One of the first exact methods in the literature to globally compare two sequences is NW [9]. It is based on dynamic programming and calculates a similarity matrix of size $m$ x $n$, where $m$ and $n$ are the sizes of the sequences. NW has time and space complexity O($mn$). The NW algorithm was modified by Smith-Waterman (SW) [14] to deal with local alignments. SW [14] is also based on dynamic programming, with quadratic time and space complexity.

Parallel processing is often used as an alternative to reduce the execution time of these exact methods. Most of the parallel strategies proposed in the literature [1][2][4][6][14] use the wavefront method to calculate the similarity matrix. In this method, the amount of parallelism is non-uniform. All of these proposals use all available processors to perform computations and, with the exception of [4], all consider homogeneous environments. Nevertheless, it has been observed that, for small sequence sizes, bad speedups are obtained when all available processors are used [1] [4] [6].

This article proposes and evaluates an application-specific task allocation framework for heterogeneous environments. Our goal is to determine which processors will be assigned to biological sequence comparison applications. To do that, our policies can take into account application-specific issues (sequence sizes, data dependency patterns) and environment-specific characteristics (processing power and communication costs). In our framework, many allocation strategies can be integrated. By now, we have implemented four policies.

The results obtained in a 10-machine heterogeneous environment show that the number of machines assigned is dependent on the size of the problem and on the allocation policy. The results obtained when comparing a hundred 1Kbp (kilo base pairs) DNA sequences with a single 1Kbp DNA sequence with a real biological sequence comparison parallel application [3] presented a reduction of 54.2% on the total execution time, when compared with the fixed policy, which assigns all available processors to the computation.

The remainder of this article is organized as follows. Section 2 describes the sequence alignment problem, presents the basic algorithms to solve it and some parallel variations. Section 3 describes the design of our task allocation framework. Some experimental results are discussed in Section 4. Finally, Section 5 concludes the paper.

## 2 Biological Sequence Comparison

To compare two sequences, we need to find the best alignment between them, which is to place one sequence above the other making clear the correspondence between similar characters [12]:

```
C A - C G G T A C
C A T C G A T - C
```

In order to measure the similarity between two sequences, a score can be calculated as follows. Given an alignment between sequences $s$ and $t$, the following values are assigned, for instance, for each column: a) +1, if both characters are identical (*match*); b) -1, if the characters are not identical (*mismatch*); and c) -2, if one of the characters is a space (*gap*). The score is the sum of all these values. The similarity between two sequences is the highest score.

The algorithm NW [9] is an exact method based on dynamic programming to obtain the best global alignment between two sequences. It is divided in two phases: create the similarity matrix and obtain the best global alignment.

The first phase receives input sequences $s$ and $t$, with $|s| = m$ and $|t| = n$, where $|s|$ represents the size of sequence $s$. The similarity matrix is denoted $A_{m+1,n+1}$, where $A_{i,j}$ contains the similarity score between prefixes $s[1..i]$ and $t[1..j]$. At the beginning, the first row and column are filled with the values -$gi$, where $i$ is the size of the non-empty subsequence and $g$ is the gap penalty. This represents the cost of aligning a non-empty subsequence with an empty one. Note that $A_{0,0} = 0$. The remaining elements of $A$ are obtained from equation 1. In equation 1, $p(i,j) = 1$ if $s(i)=t(j)$ (*match*) and $-1$ otherwise (*mismatch*). The total score between sequences $s$ and $t$ is the value contained in cell $A_{m+1,n+1}$. Note that the value of each matrix cell $A_{i,j}$ depends on $A_{i-1,j}$, $A_{i,j-1}$ and $A_{i-1,j-1}$.

$$sim(s[1..i],t[1..j]) = \max \begin{cases} sim(s[1..i],t[1..j-1]) \text{ -2} \\ sim(s[1..i-1],t[1..j-1]) \text{ +}p(i,j) \\ sim(s[1..i-1],t[1..j]) \text{ -2} \end{cases} \quad (1)$$

Figure 1 presents the similarity matrix between sequences $s$ = AGTAC and $t$ = AGTC. The arrows indicate the cell from where the value was obtained.

| | - | A | G | T | A | C |
|---|---|---|---|---|---|---|
| - | 0 | ←-2 | ←-4 | ←-6 | ←-8 | ←-10 |
| A | ↑-2 | ↖1 | ←-1 | ←-3 | ←↖-5 | ←-7 |
| G | ↑-4 | ↑-1 | ↖2 | ←0 | ←-2 | ←-4 |
| T | ↑-6 | ↑-3 | ↑0 | ↖3 | ←1 | ←-1 |
| C | ↑-8 | ↑-5 | ↑-2 | ↑1 | ↖2 | ↖2 |

**Fig. 1.** Similarity matrix to globally align two DNA sequences

To obtain the best global alignment, the algorithm starts from cell $A_{m+1,n+1}$ and follows the arrows until cell $A_{0,0}$ is reached. A left arrow in $A_{i,j}$ (figure 1) indicates the alignment of $s[i]$ with a gap in $t$. An up arrow represents the alignment of $t[j]$ with a gap in $s$. Finally, an arrow on the diagonal indicates that $s[i]$ is aligned with $t[j]$.

To obtain the similarity between parts of the sequences, local alignment must be used (SW) [14]. Like NW, SW is also based in dynamic programming with quadratic time and space complexity. However, there are three basic differences between them.

The first difference is on the initialization of the first row and column, which are filled with zeros in SW. The second difference involves the equation used to calculate the remaining cells since, in SW, no negative values are allowed. In order to do that, the value zero is included in equation 1.

The third difference concerns the cell used to start the traceback process. To obtain the best local alignment, the SW algorithm starts from the cell which has the highest value, following the arrows until the value zero is reached.

In the NW and SW algorithms, most of the time is spent calculating the similarity matrix $A$ and this is the part which is usually parallelized. The access pattern presented by the matrix calculation is non-uniform and the parallelization strategy that is traditionally used in this kind of problem is known as the wavefront method [18].

Figure 2 illustrates the wavefront method for 4 processors, where each processor calculates a subset of columns of the similarity matrix. At the beginning of the computation, only P1 is computing (figure 2.a). When P1 finishes calculating the values of a border column, it sends them to P2, that can start calculating (figure 2.b). In figure 2.c, the maximum parallelism is attained.

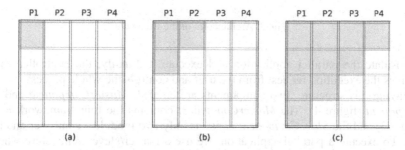

**Fig. 2.** The wavefront method

Using the wavefront method, many parallel variations of the basic algorithm have been proposed in the literature [1][2][4][14]. These proposals basically differ on the strategy used to reduce the space and/or time needed to perform computations. With the exception of [11], all use the wavefront method.

## 3   Design of the Task Allocation Framework

Our task allocation framework is designed for parallel applications that follow the wavefront pattern (figure 2) and will execute on heterogeneous environments.

The goal of our task allocation framework is to decide which processors will execute parallel tasks. Having a set of $p$ available processors and a parallel application, our framework chooses $p'$ processors ($p' \leq p$) and splits the parallel application into $p'$ tasks.

### 3.1   Overview of the Architecture

The designed framework has the modular architecture shown in figure 3. The resource discovery module is responsible for retrieving information about the heterogeneous system, generate a structured data set and store it. The task allocation module contains the task allocation policies. It is responsible to generate the processor x task map as well as to add new policies to our framework. The execution module is responsible to

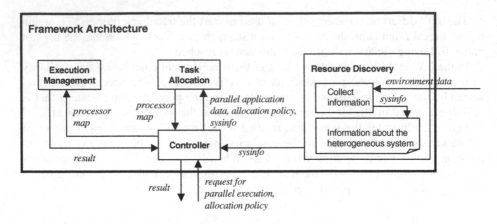

**Fig. 3.** Architecture of the framework

encapsulate the parallel application and execute it. Finally, the controller module receives the execution request from the user and controls the whole process.

Having this structure, two components are created: *MasterComponent* and *Slave-Component* (figure 4). The *MasterComponent* contains the four framework modules (figure 3) and the *SlaveComponent* contains only two modules: controller and execution. To execute a parallel application, we use a master/slave architecture where the master processor is responsible to allocate tasks to processors and also to execute a parallel task, acting as processor P1 (figure 2). Since the first processor (P1) in the wavefront method always finishes first, there is no conflict between these two roles. The other machines only execute a parallel task, communicating with the neighbors and the master processor.

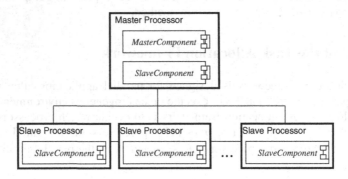

**Fig. 4.** Placement of the components in the architecture

## 3.2 Resource Discovery Module

Our resource discovery module retrieves three types of information: names of the available $p$ processors, processing power and communication cost.

The processing power is defined in the context of a parallel dynamic programming application as the cost $e$ to fill a single position of the similarity matrix (figure 1).

The communication cost is the cost of executing a *ping* application between each pair of machines $(c_{i,j})$, divided by two.

### 3.3 Task Allocation Module

The task allocation module is the core of our framework. The interface with the other modules is the function *obtainMapping (matrixDimensions, sysinfo, policy, procWeight, commWeight)*. This function is executed by the Controller component, that must provide the size of the similarity matrix *(matrixDimensions)*, information about the heterogeneous system *(sysinfo)*, the number of the policy to be used *(policy)* and the weights that must be associated with processing and computation *(procWeight* and *commWeight)*. This function returns a structured data type that represents the processor x task mapping *(map)*, which contains the size of the map, the name of the processors that will execute the parallel application and a list of begin and end columns in the similarity matrix that each processor will calculate. By now, each processor is assigned the same amount of work.

The behavior of the function *obtainMapping* is very simple, consisting of a *case* instruction that has as parameter the allocation policy *(policy)*. In order to integrate new policies to our framework, the programmer must add his/her function into this case.

In order to evaluate the generality and simplicity of our framework, we integrated 4 allocation policies to it: *fixed, ProcCom, ComProc* and *ComProcSync*.

The *fixed* policy is traditionally used to execute sequence comparison applications [11] [4] [3] [1]. Having two sequences of size $m$ and $n$ and $p$ available processors, the computation of the similarity matrix is divided among the $p$ processors, where each processor will generally calculate $m * (n/p)$ matrix cells.

The other three policies determine which $p'$ processors out of $p$ will execute tasks. They use the same idea and they basically differ on how the execution time is estimated and also on the weight assigned to processing power and communication. Figure 5 presents this basic algorithm.

```
allocationPolicy (matrixDimensions, sysinfo, procWeight, commWeight)
begin
   processor[p] = sortProcessors(sysinfo, procWeight, commWeight);
   time[1] = estimate_execution_time(processor[1]);
   for i=2 to p do
   begin
      time[i] = estimate_execution_time(processor[1..i]);
      if (time[i] < time[i-1])
         if  (i < p)
           continue_loop;
         else
           map = build_map(processor[1..p]);
      else
         begin
           map = build_map(processor[1..i-1]);
           break_loop;
         end
      return(map);
end
```

**Fig. 5.** Basic algorithm used in four allocation strategies

First, the processors are sorted according to their processing power or communication cost or a combination of both in such a way that the vector *processor* is filled with the best processor in its first position (*processor[1]*). Second, the execution time is estimated with the chosen policy for the best processor. After that, execution times are estimated using a combination of *i* processors until the estimated time for *i* processors is higher than the one for *i-1* processors. When this condition holds or when the number of available processors is attained, the loop is broken and a processor map is built with the best *i-1* (or *p*) processors.

Policies *ProcCom* and *ComProc* sort processors by processing power and communication cost, respectively. This is the only difference between them. In both cases, we assume that the execution time is limited by the processing power of the slowest processor. Also, the communication time is a function of the total amount of data transferred and the associated communication cost. Formula 2 illustrates this computation, where *i* is the number of processors under consideration, and the other parameters are explained in section 3.2.

$$t_i = e_i * (m/i) * n + n \sum_{j=1}^{i-1} c_{j,j+1} \tag{2}$$

Policy *ComProcSync* is more ellaborated than the previous ones since synchronization due to the wavefront pattern is also taken into consideration. Analyzing the parallelization of the wavefront method (figure 2), we can see that, first, only processor P1 computes. After the first communication, P1 and P2 compute in parallel. Only after the communication between P2 and P3, P3 can start computing, and so on. Using these observations, we generated formula 3 for the estimation with more than one processor. With one processor, formula 2 is used.

$$t_i = (m/i) * \sum_{j=1}^{i-1} e_j * \sum_{j=1}^{i-2} c_{j,j+1} + n * (c_{i-1,i} + e_i * (m/i)) \tag{3}$$

### 3.4 Execution Module

The goal of the execution module is to isolate the application details from the other framework modules. In this case, the *MasterComponent* (figure 4) distributes the work to the tasks and waits for the results whereas the *SlaveComponent* receives the work, does the computation and sends the results to the master.

## 4 Experimental Results

A prototype of the proposed framework was implemented in C++, using the MPI2 interface. The experimental results were collected in two research laboratories (LabPos and LaICo) interconnected by a 10Mbps switch. In our tests, we used machines with only one processor.

The hardware configuration of the ten machines used in the tests is shown in table 1 as well as the processing time to calculate one similarity matrix cell. All machines run Red Hat Linux 3.2.2 and mpich2 1.0.3.

In table 2, we can see that the LabPos machines present the best communication times. Nevertheless, communication times of pos03 and pos04 are sensibly higher than the times collected from the other machines in the same laboratory. On the other hand, the LaICo machines present the best processing times (table 1).

In our tests, we used a real bioinformatics parallel application [3] that was encapsulated by the execution module (section 3.4). In all executions, the *MasterComponent* starts processing and executes the task allocation module to create the processor x task map using the information previously retrieved by the resource discovery module (tables 1 and 2). Having the map, the *MasterComponent* dynamically instantiates the slave tasks (*SlaveComponent*) using the MPI2 spawn call.

**Table 1.** Hardware configuration and processing times of the machines used in the tests

| Lab | Machine | CPU | Memory | HD | Processing time (s) |
|---|---|---|---|---|---|
| LabPos | pos03, pos04 pos06, pos08 pos09, pos10 pos14 | AMD Athlon 1GHz | 256MB | 18GB | 0.000426 |
| LaICo | fau, magicien carbona | Pentium 4 1.7GHz | 256MB | 28GB | 0.000270 |

**Table 2.** Communication costs (s)  among the machines

|  | pos08 | pos06 | pos04 | pos03 | pos14 | pos10 | pos09 | fau | mag. | carb. |
|---|---|---|---|---|---|---|---|---|---|---|
| **pos08** | 0 | 0.086 | 0.084 | 0.072 | 0.093 | 0.094 | 0.089 | 0.193 | 0.189 | 0.188 |
| **pos06** | 0.099 | 0 | 0.071 | 0.085 | 0.078 | 0.081 | 0.080 | 0.191 | 0.187 | 0.185 |
| **pos04** | 0.086 | 0.089 | 0 | 0.085 | 0.080 | 0.082 | 0.081 | 0.192 | 0.189 | 0.185 |
| **pos03** | 0.145 | 0.168 | 0.166 | 0 | 0.186 | 0.188 | 0.187 | 0.388 | 0.373 | 0.375 |
| **pos14** | 0.176 | 0.151 | 0.144 | 0.173 | 0 | 0.172 | 0.165 | 0.389 | 0.389 | 0.377 |
| **pos10** | 0.089 | 0.077 | 0.076 | 0.087 | 0.080 | 0 | 0.084 | 0.195 | 0.191 | 0.197 |
| **pos09** | 0.092 | 0.078 | 0.073 | 0.093 | 0.081 | 0.082 | 0 | 0.185 | 0.183 | 0.189 |
| **fau** | 0.197 | 0.209 | 0.195 | 0.202 | 0.209 | 0.212 | 0.199 | 0 | 0.158 | 0.151 |
| **mag.** | 0.206 | 0.211 | 0.212 | 0.204 | 0.210 | 0.210 | 0.203 | 0.181 | 0 | 0.149 |
| **carb.** | 0.204 | 0.205 | 0.205 | 0.211 | 0.200 | 0.208 | 0.203 | 0.182 | 0.162 | 0 |

We run our prototype with the four allocation policies for the following comparisons: (a) one 1Kbp sequence is compared with a hundred 1Kbp sequences; (b) one 2Kbp sequence is compared with a hundred 2Kbp sequences and (c) one 4Kbp sequence is compared with a hundred 2Kbp sequences. The results are shown in tables 3, 4, and 5, respectively. In these tables, the comparison time is the time needed to execute the parallel application and the total execution time is the wallclock time for the whole application, including the execution of the task allocation algorithm, the spawn of the worker tasks, the application execution and the finalization procedures. The last column presents the reduction in the comparison time, when compared with the *fixed* policy.

**Table 3.** Times to compare one 1Kbp sequence with 100 1Kbp sequences

| Policy | Machines | Comparison Time (s) | Total Time (s) | Reduction |
|---|---|---|---|---|
| fixed | 10 | 4.39 | 5.56 | 0% |
| ProcCom | 3 | 2.01 | 2.77 | 54.2% |
| ComProc | 8 | 3.50 | 5.13 | 20.3% |
| ComProcSync | 5 | 3.91 | 5.17 | 11.0% |

In table 3, the best results were obtained with the *ProcCom* policy, which chose the three machines that belong to the LaICo laboratory. This happened because *ProcCom* organizes the machines by the computing power (table 1). This indicates that the parallel application is limited by the processing time and not by the communication cost, as suppose policies *ComProc* and *ComProcSync*.

**Table 4.** Times to compare one 2Kbp sequence with 100 2Kbp sequences

| Policy | Machines | Comparison Time (s) | Total Time (s) | Reduction |
|---|---|---|---|---|
| fixed | 10 | 8.09 | 10.13 | 0% |
| ProcCom | 3 | 4.75 | 5.68 | 41.3% |
| ComProc | 8 | 7.85 | 9.48 | 3.0% |
| ComProcSync | 5 | 8.18 | 9.60 | -1.1% |

In table 4, the same situation is observed since policies *ProcCom* and *ProcCom-Sync* present the best results. Policy *ComProcSync* presented the worst results but, even in this case, the total execution time was better than the one presented by the *fixed* policy. Policy *ComProc*achieved better results than *ComProcSync* since *Com-Proc* allocated all LabPos machines and one LaICo machine. This reinforces the idea that the sorting criterion must be processing time.

**Table 5.** Times to compare one 4Kbp sequence with 100 2Kbp sequences

| Policy | Machines | Comparison Time (s) | Total Time (s) | Reduction |
|---|---|---|---|---|
| fixed | 10 | 10.69 | 11.90 | 0% |
| ProcCom | 3 | 9.88 | 10.61 | 7.6% |
| ComProc | 8 | 11.313 | 12.93 | -5.8% |
| ComProcSync | 8 | 11.541 | 12.94 | -7.9% |

The scenario presented in table 5 indicates that the sequence sizes are almost big enough to justify the use of all available machines (*fixed* policy). Even in this case, policy *ProcCom* behaved better than the *fixed* policy. Policy *ComProcSync* chose 8 machines, instead of 5 (tables 3 and 4), indicating that more machines must be used since the size of the sequences is augmented.

# 5  Conclusions and Future Work

In this paper, we proposed and evaluated a task allocation framework for parallel sequence comparison applications that run on heterogeneous environments. Our framework is quite complete since it integrates resource discovery, task allocation and parallel execution. Also, allocation policies can be integrated to it in a relatively simple way.

The results obtained in an 10-machine heterogeneous environment presented a great reduction on the execution time when compared to the *fixed* allocation policy. To compare a 1Kbp DNA sequence with a hundred 1Kbp sequences, we achieved 54.2% of reduction on the comparison execution time, using 3 processors, instead of 10. As long as the sequence sizes are augmented, the policies tend to allocate more processors to the computation, having a behavior that is close to the *fixed* policy.

As future work, we intend to integrate new allocation policies and parallel sequence comparison applications to our framework. Also, we intend to evaluate our framework in a more complex environment, composed by more machines and more distinct networks. Finally, we intend to refine our resource discovery module, allowing the periodical retrieval of information.

# References

1. Batista, R.B., Melo, A.C.M.A.: Z-align: An Exact Parallel Strategy for Biological Sequence Alignment in User-Restricted Memory Space. In: Proc. of the IEEE Int. Conf. on Cluster Computing, Barcelona, IEEE Digital Library, Los Alamitos (2006)
2. Boukerche, A., Melo, A.C.M.A., Ayala-Rincon, M., Santana, T.M.: Parallel Smith-Waterman Algorithm for Local DNA Comparison in a Cluster of Workstations. In: Nikoletseas, S.E. (ed.) WEA 2005. LNCS, vol. 3503, pp. 464–475. Springer, Heidelberg (2005)
3. Boukerche, A., Melo, A.C.M.A., Sandes, E.S.F., Ayala-Rincon, M.: A Parallel Exact Algorithm to Compare Very Long Biological Sequences in Clusters of Workstations. Cluster Computing 10(2), 187–202 (2007)
4. Chen, C., Schmidt, B.: Computing large-scale alignments on a multi-cluster. In: IEEE International Conference on Cluster Computing, IEEE Digital Library, Los Alamitos (2003)
5. Cuenca, J., Gimenez, D., Martinez, J.P.: Heuristics for work distribution of a homogeneous parallel dynamic programming scheme on heterogeneous systems. In: Proc. of HeteroPar, pp. 354–360. IEEE Computer Society Press, Los Alamitos (2004)
6. Driga, A., et al.: Fastlsa: a fast, linear-space, parallel and sequential algorithm for sequence alignment. In: International Conference Parallel Processing, pp. 48–56 (2003)
7. Hwang, K., Xu, Z.: Scalable Parallel Computing: Technology, Architecture, Programming, 1st edn. McGraq-Hill, New York (1998)
8. Legrand, A., et al.: Mapping and Load Balancing of Iterative Applications. IEEE Transactions on Parallel and Distributed Systems 15(6), 546–558 (2004)
9. Needleman, S.B., Wunsch, C.D.: A general method applicable to the search for similarities in the amino acid sequence of two proteins. J. Mol. Biol. 48, 443–453 (1970)
10. Papadimitri, C.H., Steiglitz, K.: Combinatorial Optimization: Algorithms and Complexity. Dover Publications Inc., Mineola, NY (1998)

11. Rajko, S., Aluru, S.: Space and Time Optimal Parallel Sequence Alignments. IEEE Transactions on Parallel and Distributed Systems 15(2), 1070–1081 (2004)
12. Setubal, J.C., Meidanis, J.: Introduction to Computational Molecular Biology. PWS Pub. Co. (1997)
13. Sinnen, O., Sousa, L.: Communication Contention in Task Scheduling. IEEE Transactions on Parallel and Distributed Systems 16(6), 503–515 (2005)
14. Smith, T.F., Waterman, M.S.: Identification of Common molecular subsequences. Journal of Mol. Biol. 147(1), 195–197 (1981)
15. Taura, K., Chien, A.: A Heuristic Algorithm for Mapping Communicating Tasks on Heterogeneous Resources. In: Proc of the 9th Heterogeneous Computing Workshop, IEEE Computer Society Press, IEEE Digital Library, Los Alamitos (2000)

# New EST Trimming Procedure
# Applied to SUCEST Sequences

Christian Baudet[1,2] and Zanoni Dias[1,2]

[1] Institute of Computing - Unicamp - Campinas - SP - Brazil
{baudet,zanoni}@ic.unicamp.br
[2] Scylla Bioinformatics - Campinas - SP - Brasil
{christian,zanoni}@scylla.com.br

**Abstract.** In order to improve EST trimming, we proposed a new method consisting of a new set of procedures to detect regions that do not belong to the sequenced organism or have low quality or low complexity. Most trimming procedures process ESTs in a pipeline where the output of an step is adopted as the input for the following one. In our method, all artifact detection steps process the raw EST and their results are combined in the last step, which outputs the trimmed sequence. This strategy reduces the occurrence of false negatives and, additionally, has the advantage of producing better artifact composition characterization for the analyzed sequences. We evaluated our method using SUCEST [1] ESTs. Based on the results, we concluded that our method suits projects that want to produce more reliable clusters.

## 1 Introduction

An expressed sequence tag (EST) [2] is a fragment of a cDNA (complementary DNA), which is a copy of an mRNA (messenger RNA). By sequencing a cDNA, we obtain a nucleotide sequence belonging to a gene that exists in the genome and is expressed by a cell.

Artifacts are regions that do not belong to the sequenced organism or have low quality or low complexity. Their presence in the ESTs influences negatively the results of the analyzes of the data produced in the project. For example, a subsequence that has high error rates, called low quality artifact, does not guarantee that its nucleotide sequence represents the real sequence found in the organism. Therefore, it should be trimmed off, making trimming procedures an important part of the sequence analysis pipeline in an EST Sequencing Project.

Some projects, like SUCEST – Sugar Cane EST Project [3], have their own trimming procedure while others use specific trimming software as ESTprep [4] or LUCY [5]. The latter is used by TIGR - The Institute of Genomic Research.

The main objective of our work is to develop a set of trimming procedures emphasizing good clustering results. We have two previous works which presents studies that we performed to achieve this goal.

In the first one [6], we introduced the idea of using independent artifact detection steps by processing *Bos taurus* ESTs, which were sequenced by the Cattle

M.-F. Sagot and M.E.M.T. Walter (Eds.): BSB 2007, LNBI 4643, pp. 57–68, 2007.
© Springer-Verlag Berlin Heidelberg 2007

EST Project [7,8,9], with trimming procedures based on the SUCEST ones. In the second work [10], we conducted a study on slippage artifacts and, as result, we developed new algorithms to detect this artifact type.

In this work, we introduce a new set of trimming procedures that adopts the "independent artifact detection steps" strategy and incorporates results of our study about slippage artifacts. Therefore, it includes a new low quality trimming procedure that was developed through exhaustive parameter testing of two different algorithms.

We evaluated our trimming procedure set by processing SUCEST sequences and comparing the clustering of the trimmed sequences with the project's official clustering [1,3].

## 2   New Trimming Procedure Set

In order to improve the EST trimming procedure and, thus, the clustering, we built a new set of procedures to detect regions that do not belong to the sequenced organism, or have low quality.

In this new set, the identification of different types of artifacts is independently made. This means that the detection of one artifact has no effect on further detections.

This strategy, introduced in previous work [6], is distinct from the one adopted by traditional trimming procedures, which execute their steps sequentially using the output of one step as input for the next one.

The following steps are performed in our trimming procedure set: ribosomal sequence discard, low quality identification, vector identification, adapter identification, poly-A/T tail identification, slipped sequence identification and, finally, short sequences removal. Every step, except the last one, process the whole EST searching for artifacts. The last step takes the list of artifacts and processes it to identify, in the sequence, the region that will be preserved for further analysis.

### 2.1   Ribosomal Sequences Discard

Ribosomal sequences discard is performed through **BLAST** [11] of all ESTs against a database populated with ribosomal sequences of organisms that are phylogenetically close to the sequenced organism. Every sequence that shows a hit with e-value lower than or equal to $10^{-10}$ is discarded. This approach is identical to the one used in the SUCEST project [3] and can be applied because ribosomal sequences are highly conserved.

Building the database with sequences of organisms that are phylogenetically close helps the BLAST searches on the detection task. Telles and da Silva [3] built the SUCEST ribosomal database with sequences phylogenetically close to the sugarcane obtaining good results. In previous work [6], we constructed a database with ribosomal sequences originated in mammal organisms to detect *Bos taurus* ribosomal sequences (Cattle EST Project) and the results were equally great.

## 2.2   Low Quality Identification

Low quality identification is performed in two steps. In the first step, the sequence is processed by a maximum subsequence algorithm [12, Section 5.8], similar to the one implemented by phred [13,14]. We set up that algorithm using 11 as minimum quality threshold.

After the maximum subsequence being found, we use a 10 bases sliding window to search regions that have an average probability error higher than or equal to 20%. For each region found, we cut the sequence into two parts at the position of minimum quality. Each part is processed by the maximum subsequence algorithm.

The method and parameters were chosen after a exhaustive parameter testing that evaluates two different algorithms: maximum subsequence (which was our choice) and slidding window used by the SUCEST project. We compared the results of both algorithms with the ones of LUCY [5], which was run with default parameters. Our choice lies on a configuration that is a little less stringent than LUCY's configuration but that increases the chance of obtaining more BLAST hits (due to longer sequences) without harms the average sequence quality. This alternative helps the gene fiding task without compromisses the clustering quality.

## 2.3   Vector Identification

Vector identification is very simple and is performed by running the software cross_match [15] to align all sequences with the vector sequence. The alignment is made with the parameters -minmatch 12 and -minscore 20. The resulting aligned regions are considered as vector artifacts.

The parameters are similar to the ones used by Telles and da Silva, but our vector detection criteria is much more simpler than their criteria. We tested this simple alternative in previous work [6] and it showed results that were equivalent to the ones obtained by the more complex procedure.

## 2.4   Adapter Identification

Adapter detection is similar to the one developed by Telles and da Silva. It uses the software swat [15]. All sequences are aligned with the adapter sequence using the parameters -gap_init −5, -gap_ext −5, -end_gap −5, -ins_gap_ext −5, -del_gap_ext −5 and a score matrix that scores every match with 1 and every mismatch with −2. The alignments whose size is greater than or equal to the adapter size minus 4 are identified as adapter artifacts.

The number subtracted from the adapter size to obtain the alignment minimum length must be configured in accordance to the size of the adapter that was used in the project. In SUCEST, two adapters were used: one 11 and 16 bp length. If the adapter is smaller than them, the number that will be subtracted from the adapter size must be lowered to avoid false positives. On the other hand, if the adapter is greater than the SUCEST ones, the number must be increased to avoid false negatives.

## 2.5  Poly-A/T Identification

Swat is also used in poly-A/T tail search. Using the same parameters of the adapter identification step, the sequence are aligned with 500 bp sequences of "A"s or "T"s.

Alignment regions showing scores of at least 10 are considered as poly-A/T tail artifacts.

Like the vector identification phase, this step is simpler than the procedure built by Telles and da Silva and it uses the same swat parameters. Our simple procedure was also evaluated on previous work [6] and it showed equivalent results, so we decided to adopt it.

## 2.6  Slippage Identification

Usually, normal trimming procedures do not consider slippage artifacts. Caused by sequencing process problems, slippage is a region that presents an abnormal distribution of echoed bases. These echoes result from reading chromatogram regions that have many signal peaks for a single nucleotide.

Although echoed bases sometimes appear with a high background noise, signal peaks are so high that base-calling softwares assign high quality values for bases that do not exist. This phenomenon prevents the removal of these regions by trimming methods based on quality.

In a previous work [10], we conducted a detailed research to develop new algorithms to detect this type of artifact.

The slipped sequence detection is based on consecutive identical bases regions identification. If a region in the sequence comprises at least 5 consecutive identical bases, it is identified as an echo region, otherwise as a normal region. Any subsequence composed by at least 8 echoes regions that represent 25% or more of its all regions is considered a slippage artifact.

## 2.7  Short Sequence Removal

After identifying all artifacts above, our trimming procedure ends with the identification of the sequence that will be used for clustering.

All artifacts found are masked and all non-masked regions are analyzed. Every region that has size lower than 100 bases are discarded. Only the non-masked region that has the highest quality sum is preserved. If there are two regions with the same sum, the longest one is chosen. Finally, if both regions have the same size, the method chooses the one that is closer to the 5′ end.

## 3   Procedure Set Evaluation

After implementing our trimming procedure set, we performed some test to evaluate it. In Section 3.1 we present the results that were obtained after processing all SUCEST ESTs. In in Section 3.2 we show the results after clustering the trimmed sequences and we compare them with the official SUCEST clustering.

## 3.1 Trimming SUCEST Sequences

In order to evaluate our trimming procedure, we worked with the sequence set produced on the SUCEST project. This set is compound of $291,689$ ESTs, which have an average length of $829.44 \pm 182.60$ bp and an average quality of $23.15 \pm 15.71$. The execution sequence adopted in this work is pictured in Figure 1. Note that, except for the short sequences removal step, any step can be performed in any order because they are independent of the other step outputs.

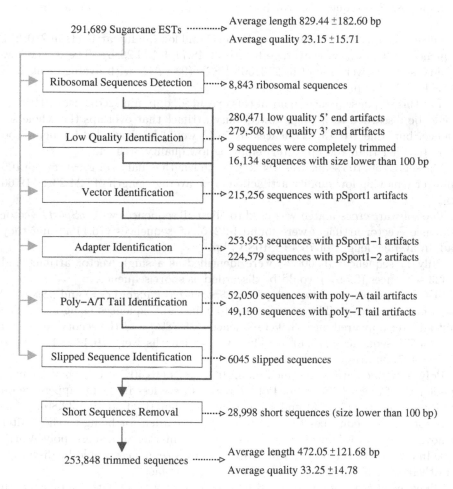

**Fig. 1.** Steps of the new EST trimming procedure. Each step is independent of the other and processes the whole sequence. The last step combines all detected artifacts and extract the region that will be used for further analysis.

To perform the ribosomal sequences discard step, we built a BLAST database with the same sequences used by Telles and da Silva. This database has the sequences GenBank AF168884 (*Zea mays* 18S rRNA), GenBank AF162215 (*Platanus occidentalis* 5.8S rRNA), and GenBank AF162215 (*Lambertua inermis* 26S rRNA).

This step produced a list of 8,843 ribosomal sequences. This number is slightly higher than the number of ribosomal sequences found by Telles and da Silva (8.473), which might be explained by the different BLAST versions used in both works. We used the version released at September $5^{th}$, 2005 (2.2.11) while they had worked with the version released at October $31^{st}$, 2000.

As the ribosomal sequences are completely discarded, we just used the remaining 282,846 sequences, which were not marked as ribosomal, in the other steps.

The low quality identification step found 5' end low quality artifacts in 280,471 sequences (99.16%), with average length of 48.71 ± 109.29 bp. The 3' end low quality artifacts were found in 279,508 ESTs (98.82%), with average length of 288.84 ± 223.25 bp.

If in this step a sequence removal was executed, only nine sequences (0.003%) would be discarded as a single low quality artifact that overlaps the whole sequence, but other 16,134 sequences (5,70%) would be discarded for being too short (smaller than 100 bp) after removing low quality artifacts.

The set of 266,703 sequences (94.29%) with length equal to or greater than 100 bp after removing low quality artifacts had an average length of 524.23 ± 119.66 bp.

The software cross-match was used to align all sequences with *pSport1* vector sequence. Vector artifacts were found in 215.265 sequences (76.11%) and they had an average length of 76.49 ± 108.15 bp.

Only 17 sequences (0,006%) were identified as a single vector artifact and 7,323 sequences (2.59%) would be discarded as short sequences.

SUCEST project used two adapters: *pSport1-1* (ccacgcgtccg) and *pSport1-2* (tcgacccacgcgtccg). They were aligned against all sequences using swat. The first adapter appeared on 253,953 sequences (89.78%) and the second appeared on 224.579 sequences (79.40%). The average lengths were 10.48 ± 1.27 bp e 15.79 ± 0.77 bp, respectively.

Poly-A artifacts were identified on 52,050 ESTs (18.40%). They had an average length of 31.45 ± 36.71 bp. Poly-T regions were less frequent, appearing on 49,130 sequences (17.37%) and showing average length of 30.61 ± 32.80 bp.

A total of 47 sequences (0.02%) would be discarded for being to short after removing poly-A tail. One of these sequences was marked as a single poly-A artifact. In the case of the poly-T tail artifacts, only 4 sequences would be discarded and there were no sequence marked as a single artifact.

Slippage artifacts appeared on 6,045 sequences (2.14%) with average length of 196.35 ± 139.19 bp. Only 293 ESTs (0.10%) would be discarded for being too short. Just one of these ESTs was a single artifact that overlapped the whole sequence.

**Table 1.** Number of artifacts and average length for each artifact type. Average length was not calculated for ribosomal artifacts because the method does not specify the artifact limits in the sequence, it just marks the sequence for discard.

| Artifact | Number of artifacts | Average Length |
|---|---|---|
| Ribossomal sequence | 8, 843 | – |
| 5′ low quality end | 280, 471 | 48.71 ± 109.29 |
| 3′ low quality end | 279, 508 | 288.84 ± 223.25 |
| pSport1 vector | 250, 705 | 76.49 ± 108.15 |
| pSport1-1 adapter | 253, 953 | 10.48 ± 1.27 |
| pSport1-2 adapter | 224, 579 | 15.79 ± 0.77 |
| Poly-A | 52, 050 | 31.45 ± 36.71 |
| Poly-T | 49, 130 | 30.61 ± 32.80 |
| Slippage | 6, 986 | 196.35 ± 139.19 |

Finally, the last step grouped all artifacts and worked to identify the sequences that would be preserved. The Table 1 shows the result of all previous steps (Number of artifacts and average artifact length).

After processing the artifact list and discarding all sequences with length smaller than 100 bp, a total of 253, 848 sequences (87, 03% of 291.689 sequeces) were preserved. The average length was 472.05 ± 121.68 and the average quality was 33.25 ± 14.78.

For the same initial set of sequences, the procedure set developed by Telles and da Silva preserved 237, 954 ESTs (81, 56%), which had an average length of 641.57 ± 139.79 bp and an average quality of 27.74 ± 14.30.

As shown, our trimming procedure set discards less sequences than the procedure used on SUCEST project. Moreover, our sequences were shorter and had higher quality than the official SUCEST trimmed sequences.

## 3.2   Clustering Trimmed Sequences

For the purpose of evaluating the quality of our trimming procedure, we decided to cluster the trimmed sequences and compare the obtained clustering with the official SUCEST one.

As the computer used to cluster the sequences had limited memory, we had to select a smaller set of sequence to perform the cluster comparison. It was done by sorting all sequences by name and selecting those that were in odd positions, resulting in, approximately, half of the sequences of each library. The 145, 845 ESTs selected to be processed and clustered have an average length of 834.64 ± 182.26 bp and average quality of 23.08 ± 15.67.

After processing this set, SUCEST trimming procedure preserved 118, 991 sequences with average size of 643.82 ± 141.32 bases and average quality of 27.69±15.39. Our set of trimming methods preserved 126, 986 ESTs with average size of 473.33 ± 121.66 and average quality of 33.25 ± 13.15. The Table 2 shows the number of sequences, average length, and average quality for the testing set and the sets of sequences processed by each method.

**Table 2.** Number of sequences, average length, and average quality for the set of selected sequences and for the sets of sequences processed by both methods (SUCEST trimming procedure and our trimming procedure)

|  | Sequences | Average Length | Average Quality |
|---|---|---|---|
| Selected sequences | 145, 845 | 834.64 ± 182.86 | 23.07 ± 14.98 |
| Processed by SUCEST methods | 118, 991 | 643.82 ± 141.32 | 27.69 ± 14.30 |
| Processed by our methods | 126, 988 | 473.32 ± 121.66 | 33.25 ± 14.78 |

The numbers of Table 2 reinforces that our trimming procedure set preserves more sequences but with smaller lengths and higher qualities when compared to the official SUCEST trimmed sequences.

After trimming the sequences, we clustered both sets of processed sequences using `cap3` [16] with default parameters. The set of sequences processed by the SUCEST methods generated a clustering (TS) with 20, 202 singletons and 16, 394 contigs (clusters with two or more sequences). The set processed with our methods generated a clustering (BD) with 22, 479 singletons and 17, 486 contigs.

We compared both clusterings by their external consistency, internal consistency, redundancy, and number of full-length clusters.

**External consistency.** External consistency evaluation searches for clustering errors that put sequences originated on the same gene in separated clusters. For this analysis, we conducted a BLAST search comparing each cluster consensus (singletons and contigs) against all other consensus sequences from the same clustering. Then, we processed the BLAST output searching for 200 bp long alignment, with 75% minimum identity and located at maximum 10 bp far from one of the consensus extremities. After that, for each clustering we computed the number of overlaps that meet that criteria and divided it by $O(n) = n(n-1)/2$, that is the maximum number of possible overlaps for $n$ clusters.

Clustering TS showed 1, 098 overlaps that are equivalent to $1.64 \times 10^{-4}\%$ of the possible overlaps. Clustering BD showed 1, 269 overlaps and they correspond to $1.59 \times 10^{-4}\%$ of the possible overlaps. We plotted the results of this analysis in the graph of the Figure 2. Each point of this graph is result of the function $f(x) = x \times [n(n-1)/2] \times 100$, where $x$ is the number of overlaps found with $x\%$ of identity.

Proportionally to the number of clusters of each clustering, the external consistency analysis showed that the clustering BD has a number of clusters overlaps lower than that shown by clustering TS. This could indicate that our clustering separates clusters that, in fact, might have been assembled together with a frequency lower than that one of clustering TS.

**Internal consistency.** Opposed to the external consistency evaluation, the internal consistency analysis searches for clustering errors that make sequences to be grouped in the same cluster, but they should have been placed in different clusters. For this evaluation, we collected data about the discrepant bases found in contigs. A discrepant base is the one that differs from the consensus bases in the same alignment column.

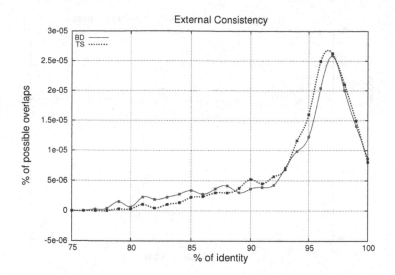

**Fig. 2.** Distribution of overlaps found in the BLAST of "all against all" clusters of each clustering [TS (dashed line) and BD (solid line)]. We considered only alignments that were at least 200 bp long, with a minimum identity of 75%, and located at maximum 10 bp far from one of the consensus ends. The $y$ axis indicates the percentage of possible overlaps that were found in function of the overlap identity. The total number of possible overlaps is given by $n(n-1)/2$, where $n$ is the total number of clusters.

For each clustered sequence we calculated its discrepant base percentage. Then we evaluated the distribution of the sequences in function of their discrepancy. The results of this analysis is plotted in the graph of Figure 3. The $y$ axis represents the percentage of discrepant in the group of clustered sequences with $x\%$ of discrepant bases.

The graph of the Figure 3 exhibits that clustering BD has a greater percentage of sequences with less than 2% of discrepant bases than clustering TS. It also shows that clustering TS has more sequences with higher discrepancy levels.

**Redundancy.** The redundancy analysis was performed to extract one more parameter for the clustering quality comparison. A high redundancy between clusters might indicate that some of them should have been grouped together, instead of being separated due to sequence quality.

To perform this analysis, every singleton and contig consensus sequences were compared to each other ("all against all") using cross_match. The objective was to identify which sequences would be grouped into "contigs" and which would stay alone as "singletons". We perform this test using two set of parameters and criteria.

The first configuration was based on the one used in the SUCEST data analysis [1]. Cross_match was ran with parameters -penalty $-10$, -minmatch 32, and -minscore 77. Every alignment of sequences, independently of its length, was considered to group the sequences in "contigs".

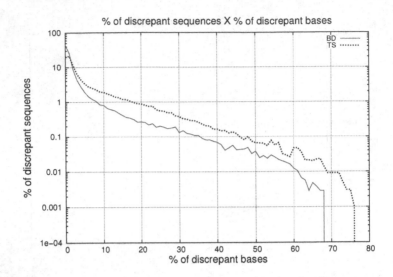

**Fig. 3.** Distribution of discrepant sequences in function of their discrepant bases percentage. For each sequence that was participating of a cluster with size two or more ESTs, we calculated its discrepant base (base that do not agree with the consensus base in the same alignment column) percentage.

Using this criteria, clustering TS showed a redundancy level of 18.56% (25, 902 "singletons" e 3, 903 "contigs"), while clustering BD presented a redundancy level of 14.51% (30, 749 "singletons" e 3, 417 "contigs").

In the second configuration, cross_match was ran with default parameters and only alignments with at least 100 bp length and 98% of identity were considered to be part of the "contigs".

This configuration pointed out that clusterings TS and BD have redundancy levels of 6.10% (32, 818 "singletons" e 1, 547 "contigs") and 5.97% (35, 989 "singletons" e 1, 592 "contigs"), respectively.

Both redundancy tests brought out that clustering TS has higher redundancy level than clustering BD. This is particularly interessant considering that clustering BD has more sequences than clustering TS.

These results evidence that clustering TS may have more clustering errors caused by its sequence average quality, which is lower than the clustering BD.

**Full-length Clusters.** Finally, the last test was performed to evaluate the amount of full-length clusters that can be found in each clustering. The process adopted to identify this kind of cluster was the same executed in the SUCEST data analysis [1].

A full-length cluster is the one that besides having at least one BLAST hit against nr with e-value lower than or equal to $10^{-40}$, have an alignment starting in the first 15 positions of the subject sequence.

Clustering TS exhibited 4, 941 (13.50%) full-length clusters. Decompounding this set, 3, 702 full-length clusters were found in contigs and 1, 239 were found in singletons.

Clustering BD had less full-length clusters, with 3, 526 found in contigs and 882 in singletons, totalizing 4, 408 (11.03%) full-length clusters. This result may be a consequence of smaller average size of the sequences processed by our methods

## 4   Conclusion

During this work we developed a set of trimming procedures to achieve simplicity and independence among steps.

Simplicity is related to the easiness of implementation and maintenance. Independence among steps is linked to the easiness of configuration (you can turn on/turn off any step without cause problems to any other) and, mainly, to the possibility of trimming error reduction.

When one step depends on the output of the previous one, its results can be influenced by an incorrect artifact identification. On the other hand, a trimming method that works with the whole sequence, is not affected by this problem.

The results showed that every step of our trimming procedure was able to identify its target artifact. The clustering built with the sequences processed by it exhibited better external and internal consistency as well as lower redundancy level than the resulting clustering using the sequences produced by the SUCEST official trimming procedure.

The main difference between the two trimming sets is explained by our method having a more stringent low quality trimming, making the clusters more reliable, despite less full-length clusters.

Based on these analysis, we conclude that our trimming procedure suits projects that aim to produce more reliable clusters. If the objective is to discover full-length clusters for gene annotation, we recommend a reduction of the low quality trimming stringency.

## Acknowledgments

The present study was developed at Scylla Bioinformática (www.scylla.com.br) and was partially supported by FAPESP (Fundação de Amparo à Pesquisa do Estado de São Paulo – www.fapesp.br), grant numbers 2003/07748-9 and 2004/09417-2.

## References

1. Vettore, A.L., da Silva, F.R., Kemper, E.L., Arruda, P.: The libraries that made SUCEST. Genetics and Molecular Biology 406, 151–157 (2001)
2. Adams, M.D., Kelley, J.M., Gocayne, J.D., Dubnick, M., Polymeropoulos, M.H., Xiao¡ H., Merril, C.R., Wu, A., Olde, B., Moreno, R.F., Kerlavage, A.R., McCombie, W.R., Venter, J.C.: Complementary DNA Sequencing: Expressed Sequence Tags and Human Genome Project. Science 252, 1651–1656 (1991)

3. Telles, G.P., da Silva, F.R.: Trimming and clustering sugarcane ESTs. Genetics and Molecular Biology 24(1-4), 17–23 (2001)
4. Scheetz, T.E., Trivedi, N., Roberts, C.A., Kucaba, T., Berger, B., Robinson, N.L., Birkett, C.L., Gavin, A.J., O'Leary, B., Braun, T.A., Bonaldo, M.F., Robinson, H.P., Sheffield, V.C., Soares, M.B., Casavant, T.L.: ESTprep: preprocessing cDNA sequence. Bioinformatics 19(11), 1318–1324 (2003)
5. Chou, H., Holmes, M.H.: DNA sequence quality trimming and vector removal. Bioinformatics 17, 1093–1104 (2001)
6. Baudet, C., Dias, Z.: New EST Trimming Strategy. In: Setubal, J.C., Verjovski-Almeida, S. (eds.) BSB 2005. LNCS (LNBI), vol. 3594, pp. 206–209. Springer, Heidelberg (2005)
7. Band, M.R., Larson, J.H., Rebeiz, M., Green, C.A., Heyen, D.W., Donovan, J., Windish, R., Steining, C., Mahyuddin, P., Womack, J.E., Lewin, H.A.: An Ordered Comparative Map of the Cattle and Human Genomes. Genome Research 10, 1359–1368 (2000)
8. Cattle EST Project: The W. M. Keck Center for Comparative and Functional Genomics, University of Illinois at Urbana-Champaign (January 2005), http://titan.biotec.uiuc.edu/cattle/cattle_project.htm
9. Ma, R.Z., van Eijk, M.J.T., Beever, J.E., Guérin, G., Mummery, C.L., Lewin, H.A.: Comparative analysis of 82 expressed sequence tags from a cattle ovary cDNA library. Mammalian Genome 9, 545–549 (1998)
10. Baudet, C., Dias, Z.: Analysis of slipped sequences in EST projects. Genetics and Molecular Research 5(1), 169–181 (2006)
11. Altschul, S.F., Gish, W., Miller, W., Myers, E.W., Lipman, D.J.: Basic local alignment search tool. Journal of Molecular Biology 215(3), 403–410 (1990)
12. Manber, U.: Introduction to Algorithms. Addison-Wesley, Reading (1989)
13. Ewing, B., Hillier, L., Wendl, M.C., Green, P.: Base-Calling of Automated Sequencer Traces Using Phred. I. Accuracy Assessment. Genome Research 8(3), 175–185 (1998)
14. Ewing, B., Green, P.: Base-Calling of Automated Sequencer Traces Using Phred. II. Error Probabilities. Genome Research 8(3), 186–194 (1998)
15. Green, P.: Phrap Homepage: phred, phrap, consed, swat, cross_match and Repeat-Masker Documentation (March 2004), http://www.phrap.org
16. Huang, X., Madan, A.: CAP3: a DNA sequence assembly program. Genome Research 9, 868–877 (1999)

# A Method for Inferring Biological Functions Using Homologous Genes Among Three Genomes

Daniel A. S. Anjos, Gustavo G. Zerlotini, Guilherme A. Pinto,
Maria Emilia M. T. Walter, Marcelo M. Brigido, Guilherme P. Telles,
Carlos Juliano M. Viana, and Nalvo F. Almeida

[1]Department of Computer Science, University of Brasília, Brasília, Brazil
[2]Institute of Biology, University of Brasília, Brasília, Brazil
[3]Institute of Mathematical Sciences and Computing, University of São Paulo,
São Carlos, Brazil
[4]Department of Computer Science and Statistics, Federal University of Mato Grosso
do Sul, Campo Grande, Brazil
{daniels,gustavogz,mariaemilia,brigido}@unb.br, gap@cic.unb.br,
gpt@icmc.usp.br, carlosjmviana@gmail.com, nalvo@dct.ufms.br

**Abstract.** In this work, we propose $n3GC$, a method to infer a particular biological function in an organism, by finding homologous genes among three genomes, comparing the genes of the investigated organism with the genes of two other organisms, one having and the other not having this function. Our $n3GC$ method takes as input *previously* identified families of paralogous genes in each one of the genomes, and produces a three set Venn diagram, each set representing a genome. The intersection of three (two) sets shows the families of similar genes having strong similarities among the three (two) genomes. The gene families of a genome not having strong similarities with any family of the other two genomes appear outside the intersections. We have used our method to determine potential pathogenic genes of the *Paracoccidioides brasiliensis* fungus, comparing it with seven fungi, three at a time, one pathogenic and the other non-pathogenic. To validate $n3GC$, we first investigate the Pfam classification of the families belonging to the intersections and compare with INPARANOID and $3GC$ methods.

## 1 Introduction

The great volume of genomic information generated by many sequencing genome projects around the world allowed researchers to infer biological functions of an organism based on already known biological functions of phylogenetic close organisms. Strong similarities among genes of different genomes indicate, to biologists, genes that could participate in specific cellular processes.

Generally, in order to infer related functions, researchers develop methods to obtain paralogous and orthologous genes among genomes. Some of them identify orthology relationships by building or analyzing phylogenetic trees, but these

M.-F. Sagot and M.E.M.T. Walter (Eds.): BSB 2007, LNBI 4643, pp. 69–80, 2007.
© Springer-Verlag Berlin Heidelberg 2007

methods require a great volume of computational resources, and are difficult to implement [10]. Other methods are based on all-against-all gene comparisons that are easier to implement and present good results [13,5,12,7,9,4,11,14]. Some methods combine phylogeny and comparative genomics [6]. These previous methods may in principle be used to identify genes potentially involved in a specific cellular process, but they would require subsequent processing using more computational efforts. Instead, a method could be designed using three genomes directly, the investigated genome and two other phylogenetically close genomes, one having and the other not having the desired cellular process.

The objective of this article is to present $n3GC$, a method to infer a biological function by obtaining homologous genes among three genomes simultaneously. Our method takes as input *previously* identified families of similar genes in each one of the genomes, and produces a three set Venn diagram, as illustrated in Figure 1, each set representing a genome. The intersection of three (two) sets shows the families of similar genes having strong similarities among the three (two) genomes. The gene families of a genome not having strong similarities with any family of the other two genomes appear outside the intersections. The idea of producing a Venn diagram comparing three genomes was introduced by Telles and co-authors [15] but the method used to obtain the diagram is different and does not consider paralogy relationships inside one genome. We used our method to determine potential pathogenic genes of the *Paracoccidioides brasiliensis* fungus, comparing it with seven fungi, three at a time, one pathogenic and the other non-pathogenic. To validate $n3GC$, we investigate the Pfam classification of the families appearing in the intersections and compare these results with INPARANOID and $3GC$ methods, obtaining good results.

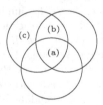

**Fig. 1.** Venn diagram representing families with homologous genes (a) among three genomes, (b) among two genomes, and (c) exclusive to one genome

In Section 2, we briefly describe the EGG method to find paralogs, that was used to produce the families of similar genes inside one genome. Afterwards, we present the $3GC$ method to find similar genes among three genomes. In Section 3 we devise the $n3GC$ method to identify homologous genes among three genomes. In Section 4, we show how we used our method to determine potential pathogenic genes of the *Paracoccidioides brasiliensis* fungus. Then we investigate the produced families classifying the genes of an intersection according to their Pfam [3] classification, and counting how many different Pfam classes appear in each intersection. At last, we compare our results with INPARANOID and $3GC$ methods. Finally, in Section 5 we conclude and suggest future work.

## 2    EGG and 3GC Methods

First we describe the *EGG* method [1], that uses BLAST [2] scores of aligned genes inside a genome to determine the families of paralogous genes. Informally, we say that two genes are paralogous using some rule involving similarity and alignment coverage. After defining initially when two genes are paralogous, the most natural way of determining the families would be considering maximal sets of pairwise paralogous genes. However, this could potentially leave outside a family many genes which are paralogous to some but not to all paralogs of that family. The *EGG* method, then, relaxes the paralogy rule to aggregate these genes to the families.

Formally, two genes $g$ and $h$ are *paralogous* if, in the alignment of $g$ and $h$ (and of $h$ and $g$): the *e-value* produced by BLAST is less than or equal to some threshold $S$; and at least $P\%$ of $|g|$ and $P\%$ of $|h|$ are covered by the alignment, for some threshold $P$, where $|g|$ is the number of residues of $g$.

The method uses a simple graph $G = (V, E)$, having a non-empty set of vertices $V$ and a set of edges $E$ to model the paralogy relationship. Each vertex $v \in V$ represents a gene $g_v$, and each edge $(u, v) \in E$ links two paralogous genes $g_u$ and $g_v$, according to the definition above. Besides $G$, the *EGG* method computes another graph $G' = (V', E')$ with $V' = V$ and $E'$ obtained with weaker thresholds $S' \geq S$ and $P' \leq P$.

The algorithm proceeds in three steps. On step 1, it finds all maximal cliques in $G$, that is, the maximal sets of pairwise paralogous genes [1]. Let $C_1, C_2, \ldots, C_t$ denote these sets. On step 2, it possibly adds genes (called *aggregate genes*) to the sets found on step 1. A gene will be an aggregate to $C_i$ if it does not belong to $C_i$ and is paralogous to at least one gene of $C_i$, using now the weaker graph $G'$. Clearly, a gene may be added to more than one set. Then, on step 3, the method breaks these ties by removing an aggregate gene from all sets to which it was added, except the one to which it has the higher identity. This is obtained from the average value of similarity between the aggregate gene and all the other genes of the set, according to the corresponding *e-value* and the alignment coverage.

The resulting sets $C_1', C_2', \ldots, C_t'$ of paralogs will be families given as input to our *n3GC* method, detailed in the next section. Before proceeding, we now give a brief description of the *3GC* method.

This previous method [15] computes a Venn diagram, as in our case, for three genomes. However, it attempts to identify genes common to three or two genomes by direct comparison between them, without considering paralogy inside each genome. The method uses BLAST expectation to determine similarity between genes of two different genomes. Then, using the average percent coverage produced by BLAST, it decides where each gene is to be placed in the Venn diagram by considering: first, sets of three pairwise similar genes, which are placed in the intersection of the three sets; then, pairs of similar genes, placed in the intersections between two sets; and, finally, the remaining genes exclusive to one genome.

Although presenting good results, this method may exhibit two disadvantages. First, it may place many paralogous genes in the intersections but without any information identifying them as paralogs. Second, for instance, it may place pairs of genes in the intersections between two genomes, whereas they should, from the biological point of view, be better placed in the intersection of the three genomes. This situation is illustrated in Figure 2, where the *best hit* similarity relation determined by the method will actually prevent the set of genes $\{g_1, g_2, \ldots, g_6\}$ from being recognized as common to the three genomes, a result which is expected if the paralogy relationship inside each genome is considered. Proper identification of these potential situations *after* the application of this method would require all-to-all gene comparisons between many combinations of intersections. Since this identification is relevant to the purpose of inferring biological functions, our method $n3GC$ shows how this can be done by identifying paralogy *before* comparison among genomes.

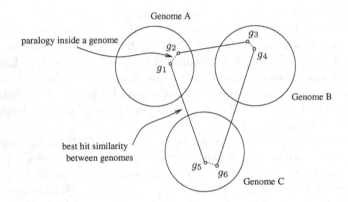

**Fig. 2.** A biological situation not considered by the previous 3GC method

## 3   The $n3GC$ Method

First we define a *hit* as the result of the comparison to detect similarities between two genes. We say that two genes $g$ and $g'$ *make a bi-directional best hit (BBH)* if $g$ is the best hit of the comparisons among $g'$ and all the genes of the set containing $g$ and vice-versa. Considering two genomes $\mathcal{G}$ and $\mathcal{G}'$, we determine if a gene $g$ of $\mathcal{G}$ is similar to a gene $g'$ of $\mathcal{G}'$ if $g$ and $g'$ make a *BBH* and both alignments have *e-value* $\leq S$ and $P\%$ of alignment coverage. The thresholds $S$ and $P$ are fixed, analogously to the steps described to identify paralogy between two genes.

Before proposing the $n3GC$ method, we make some definitions and notations. We denote a *gene* $g$ by lower case letters, and a *family of paralogous genes* $F$, by capital letters. For example, $g \in F$ means that gene $g$ belongs to family $F$. A graph $G = (V, E)$ is *tripartite* if $V$ can be partitioned into three disjoint sets $V_i$, $i \in \{1, 2, 3\}$, with no edge linking any two vertices of the same partition.

Now we detail our method. Its input is formed by the families of paralogous genes of three organisms. The families of genome $\mathcal{G}^j$ is represented by $F_i^j$, $1 \leq j \leq 3$, and $1 \leq i \leq t_j$, where $t_j$ is the number of families in $\mathcal{G}^j$.

First, we use two tripartite graphs $G$ and $G_f$ to represent, respectively, the relationship among the genes and among the families of paralogous genes of three genomes. In the graph $G = (V, E)$, we represent each gene of genome $\mathcal{G}^j$ by a vertex $g$ belonging to partition $V_j \subset V$, $1 \leq j \leq 3$. There is an edge $(g_1, g_2) \in E$ between genes $g_1 \in V_i$ and $g_2 \in V_j$, $i \neq j$, if $g_1$ and $g_2$ make *BBH*. In the experiments, we used *BBH* with thresholds $S = 10^{-5}$, $P \geq 60$ for proteins and $P \geq 0$ for translated ESTs. In the graph $G_f = (V_f, E_f)$, we represent each family by a vertex $F$ of the partition $W_j \subset V_f$, $1 \leq j \leq 3$. Each edge $(F_1, F_2) \in E_f$ links families $F_1 \in W_i$ and $F_2 \in W_j$, $i \neq j$, if there are two genes $g_1 \in F_1$ and $g_2 \in F_2$ making *BBH*.

We represent the results of the comparisons among the three partitions by *perfect triangle*[1], *triangle*, *edge* and *node*. We define a *perfect triangle* on the $G_f$ graph (Figure 3 (a)) as a set of three families $F_1^1 \in W_1$, $F_1^2 \in W_2$ and $F_1^3 \in W_3$, $W_1, W_2, W_3 \subset V_f$, having genes $g_1 \in F_1^1$, $g_2 \in F_1^2$ and $g_3 \in F_1^3$ such that $g_1$ makes BBH with $g_2$, $g_2$ makes BBH with $g_3$ and $g_3$ makes BBH with $g_1$. We define a *triangle* on the $G_f$ graph (Figure 3 (b)) as a set of three families $F_2^1 \in W_1$, $F_2^2 \in W_2$ and $F_2^3 \in W_3$, having genes $g_4, g_4' \in F_2^1$, $g_5, g_5' \in F_2^2$ and $g_6, g_6' \in F_2^3$ such that $g_4$ makes BBH with $g_5'$, $g_5$ makes BBH with $g_6$ and $g_6'$ makes BBH with $g_4'$. Note that possibly $g_i = g_i'$, for $i \in \{4, 5, 6\}$. We define an *edge* $(F_3^1, F_3^2)$, $F_3^1 \in W_1$ and $F_3^2 \in W_2$ on the $G_f$ graph (Figure 3 (c)) if there are two genes $g_7 \in F_3^1$ and $g_8 \in F_3^2$ such that $g_7$ and $g_8$ make BBH. Our method finds *perfect triangles*, *triangles* and *edges*, following this order. A family not involved in any *perfect triangle*, *triangle* or *edge* is a *node* (Figure 3 (d)).

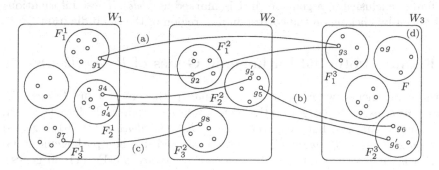

**Fig. 3.** A tripartite graph with each partition marked by a square. Families are shown inside circles. (a) A *perfect triangle*. (b) A *triangle*. (c) An *edge*. (d) A node.

After building graphs $G$ and $G_f$, we use a greedy approach to find triangles and edges corresponding to families having genes making BBH among three or two families, respectively. A depth first search until depth 3 is made first on the $G$ graph from each gene of each partition to find *perfect triangles*, and after in each family of each partition of the $G_f$ graph to find *triangles*, in order to check whether the search comes back to the initial vertex or family, as shown below.

---

[1] Our definition is different from the COG[14] triangle definition.

**Method** $n3GC$

**Input:** $F_1^1, F_2^1, \ldots, F_{t_1}^1,\ F_1^2, F_2^2, \ldots, F_{t_2}^2,\ F_1^3, F_2^3, \ldots, F_{t_3}^3$, such that each $F_i^j$,
   $j \in \{1, 2, 3\}$ and $1 \leq i \leq t_j$, is a paralogs family, of each one of three organisms.
**Output:** *perfect triangles, triangles, edges* and *nodes*
 1: {Find homologs among the three organisms}
    Find all *perfect triangles* on $G_f$, removing these families from further analysis.
    The removal is made "marking" the families belonging to the perfect triangles.
 2: Find all *triangles* on $G_f$, removing these families from further analysis. The
    removal is made "marking" the families belonging to the triangles.
 3: {Find homologs among two organisms}
    Find all *edges* on $G_f$, removing these families from further analysis. The removal
    is made "marking" the families belonging to the edges.
 4: {Find genes belonging to an organism}
    Families not marked as *perfect triangles, triangles* or *edges* on $G_f$ are considered
    *nodes*.

   The Venn diagram is generated by associating the total number of perfect triangles and triangles to the intersection of the three sets, the total number of edges to the intersection of the two corresponding sets, and the total number of genes exclusive to one genome to the corresponding set. Lists of homologs among the three organisms (triangles), homologs between two organisms (edges), and genes exclusive to one organism (nodes) are also generated. These lists are stored on files, and show the families involved in the homology relationships and the genes used to determine homology among three families (if it belongs to a *perfect triangle* or a *triangle*), homology between two families (if it belongs to an *edge*) or a family exclusive to a genome (if it is marked as *node*). These informations can be seen by clicking on the corresponding region of the Venn diagram.

## 4    Finding Potential Pathogenic Genes of *P. brasiliensis*

On the experiments, we compared the human pathogenic fungus *Paracoccidioides brasiliensis*-Pb (6022 putative genes inferred from EST data [8]) with seven fungi, being two human pathogenic - *Candida albicans*-Ca (6165 genes) and *Cryptococcus neoformans*-Cn (6578 genes), and five non-pathogenic - *Aspergillus nidulans*-An (9541 genes), *Fusarium graminearum*-Fg (11640 genes), *Magnaporthe grisea*-Mg (11109 genes), *Neurospora crassa*-Nc (10082 genes), and *Saccharomyces cerevisiae*-Sc (6305 genes). These comparisons can be used to investigate genes related to pathogenecity of the *P. brasiliensis*, a dimorphic fungus that causes a prevalent mycosis in Central and Latin America. We compared the *P. brasiliensis* putative genes with a pathogenic fungus and a non-pathogenic fungus, which generated 10 comparisons among three genes. This work is part of the *Project Pb Genome*, that sequenced 19, 718 ESTs, which generated 6, 022 genes of the *P. brasiliensis*. Now genes involved in the pathogenecity of this fungus are under investigation. Due to the lack of genomic data we used the assembled EST groups of the *P. brasiliensis* fungus.

The Venn diagrams shown in Figure 4 (the numbers indicate families) report the results of the comparisons among *P. brasiliensis*, each one of the pathogenic fungi, *C. albicans* or *C. neoformans*, and the other five non-pathogenic fungi [16].

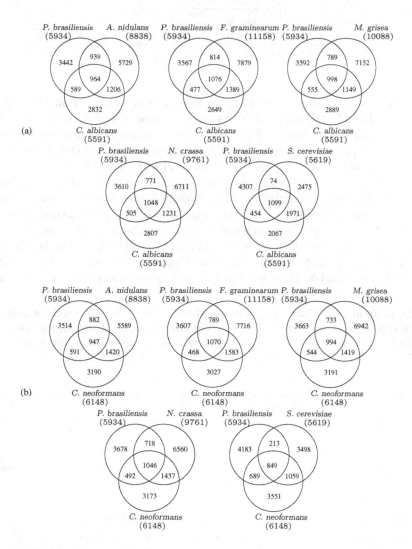

**Fig. 4.** (a) *P. brasiliensis*, *C. albicans* and non-pathogenic fungi comparisons. (b) *P. brasiliensis*, *C. neoformans* and non-pathogenic fungi comparisons.

Table 1 summarizes the execution times. These experiments were developed on a PC Pentium 4 1.5GHz with 512MB memory.

**Table 1.** Running times to identify homologs among *P. brasiliensis*, a pathogenic fungus (Pf) (in each column) and a non-pathogenic fungus (in each line)

| Time | Ca | Cn |
|---|---|---|
| Pb x Pf x An | 8m30s | 8m31s |
| Pb x Pf x Fg | 9m50s | 9m57s |
| Pb x Pf x Mg | 9m22s | 9m47s |
| Pb x Pf x Nc | 8m56s | 9m05s |
| Pb x Pf x Sc | 6m18s | 6m55s |

In order to validate our method, we first computed the Pfam [3] functions of the families identified by *n3GC* as triangles and edges. Pfam [3] is a database of multiple alignments of *proteic domains* families. A proteic domain is a region of a protein having a specific biological function. Pfam database was chosen because it classifies a gene according to its biological functions. The families forming triangles and edges were investigated to verify if the corresponding genes had the same Pfam classification. This indicates whether members of the families identified by *n3GC* share the same protein family signature. The lower the number of triangles and edges with different Pfam results, the better the *n3GC* performance. In this case, *n3GC* method would be grouping families with similar biological functions. The same methodology was used by Telles and co-authors [15] to validate their method to find homologs among three genomes.

**Table 2.** Number of Pfam results compared with the number of triangles among *P. brasiliensis*, *C. albicans* and *S. cerevisiae* genomes (columns 2 and 3), and number of triangles among *P. brasiliensis*, *C. neoformans* and *S. cerevisiae* genomes (columns 4 and 5).

| Pfam results | triangles of Pb, Ca and Sc | % | triangles of Pb, Cn and Sc | % |
|---|---|---|---|---|
| 1 | 470 | 42,8 | 371 | 43,7 |
| 2 | 523 | 47,6 | 465 | 54,8 |
| 3 | 97 | 8,8 | 9 | 1,1 |
| 4 | 7 | 0,6 | 3 | 0,3 |
| 5 | 1 | 0,1 | 0 | 0 |
| 6 | 1 | 0,1 | 1 | 0,1 |
| **Total** | 1099 | 100 | 849 | 100 |

Table 2 presents the number of Pfam results and the number of triangles among *P. brasiliensis*, *C. albicans* and *S. cerevisiae* (columns 2 and 3). Note that $90,4\%$ of the triangles among these genomes present 1 or 2 Pfam results, which indicates that the genes grouped on triangles have the same (42%) or share two Pfam functions (47%). This table also presents the results of the Pfam results and the number of triangles among *P. brasiliensis*, *C. neoformans* and *S. cerevisiae* (columns 2 and 3). We can see that 98% of the indicated triangles present 1 or 2 different Pfam results, being 43% with the same function and 54% having two similar functions.

Table 3 shows the number of Pfam results and the number of edges obtained by *n3GC* among *P. brasiliensis*, *C. albicans* and *S. cerevisiae*. Note that 97, 4% of the edges between *P. brasiliensis* and *C. albicans* presented 1 or 2 Pfam results, a very good result which indicates that 33% of the edges identified by our method has exactly the same biological function and 64% has two similar functions. Comparing *P. brasiliensis* and *S. cerevisiae* we can see that all edges have 1 (20%) or 2 (80%) biological functions. At last, for *C. albicans* and *S. cerevisiae* genomes, *n3GC* produced 99% of the edges having at most two similar functions, being 76% with the same function and 23% with two similar functions.

**Table 3.** Number of edges by number of Pfam results *P. brasiliensis*, *C. albicans* and *S. cerevisiae*

| Pfam results | Pb and Ca | % | Pb and Sc | % | Ca and Sc | % |
|---|---|---|---|---|---|---|
| 1 | 151 | 33,3 | 15 | 20,3 | 1502 | 76,21 |
| 2 | 291 | 64,1 | 59 | 79,7 | 454 | 23,03 |
| 3 | 8 | 1,8 | 0 | 0 | 10 | 0,51 |
| 4 | 2 | 0,4 | 0 | 0 | 3 | 0,15 |
| 6 | 0 | 0 | 0 | 0 | 1 | 0,05 |
| 7 | 0 | 0 | 0 | 0 | 1 | 0,05 |
| 14 | 1 | 0,2 | 0 | 0 | 0 | 0 |
| 19 | 1 | 0,2 | 0 | 0 | 0 | 0 |
| **Total** | 454 | 100 | 74 | 100 | 1971 | 100 |

**Table 4.** Number of different Pfam results and number of edges between *P. brasiliensis* and *C. neoformans*, *P. brasiliensis* and *S. cerevisiae*, and *C. neoformans* and *S. cerevisiae*

| Pfam results | Pb and Cn | % | Pb and Sc | % | Cn and Sc | % |
|---|---|---|---|---|---|---|
| 1 | 685 | 99,4 | 70 | 32,8 | 1056 | 99,7 |
| 2 | 2 | 0,3 | 141 | 66,2 | 3 | 0,3 |
| 3 | 0 | 0 | 1 | 0,5 | 0 | 0 |
| 4 | 0 | 0 | 1 | 0,5 | 0 | 0 |
| 5 | 1 | 0,15 | 0 | 0 | 0 | 0 |
| 6 | 1 | 0,15 | 0 | 0 | 0 | 0 |
| **Total** | 689 | 100 | 213 | 100 | 1059 | 100 |

Table 4 shows the number of different Pfam results and the number of edges identified by *n3GC* among *P. brasiliensis*, *C. neoformans* and *S. cerevisiae*. We can note that 99, 4% of the edges between *P. brasiliensis* and *C. neoformans* presented only 1 Pfam result. The comparisons of *P. brasiliensis* and *S. cerevisiae* indicated 99% of 1 (33%) or 2 (66%) Pfam results. *C. neoformans* and *S. cerevisiae* presented almost all genes (99, 7%) with the same function. These results suggest that our method correctly indicates groups of homologs.

Following, we compared some *n3GC* results with INPARANOID [13] (Tables 5 and 6). In this table, the column *coherent groups* shows the number of INPARANOID orthologous groups entirely contained within *n3GC* groups.

At last, we compared our results with the results of the our previous *3GC* method [15] (Tables 7 and 8). In these tables, the column *coherent groups* shows the number of *3GC* orthologous groups entirely contained within *n3GC* groups. We observe that *n3GC* joined paralogous genes separated before by *3GC*.

**Table 5.** Comparisons of the ortholog groups identified by INPARANOID and the homologous families identified on the intersection of two genomes when $n3GC$ is applied on *P. brasiliensis, C. albicans* and *S. cerevisiae* genomes

| Genomes | Assembled ESTs or proteins | $n3GC$ homologs | % | INPARANOID homologs | % | Identical groups | % | Coherent groups | % |
|---|---|---|---|---|---|---|---|---|---|
| Pb x Ca | 12037 | 3106 | 25,80 | 1474 | 12,25 | 1048 | 71,10 | 1056 | 71,64 |
| Ca x Sc | 12320 | 6140 | 49,84 | 7714 | 62,61 | 6024 | 78,09 | 6096 | 79,03 |
| Sc x Pb | 12327 | 2346 | 19,03 | 1378 | 11,18 | 84 | 6,10 | 854 | 61,97 |

**Table 6.** Comparisons of the ortholog groups identified by INPARANOID and the homologous families identified on the intersection of two genomes when $n3GC$ is applied on *P. brasiliensis, C. neoformans* and *S. cerevisiae* genomes

| Genomes | Assembled ESTs or proteins | $n3GC$ homologs | % | INPARANOID homologs | % | Identical groups | % | Coherent groups | % |
|---|---|---|---|---|---|---|---|---|---|
| Pb x Cn | 12600 | 3076 | 24,41 | 1468 | 11,65 | 1060 | 72,21 | 1064 | 72,48 |
| Cn x Sc | 12883 | 3816 | 29,62 | 5804 | 45,05 | 3690 | 63,58 | 3780 | 65,12 |
| Sc x Pb | 12327 | 2124 | 17,23 | 1378 | 11,18 | 144 | 10,45 | 778 | 56,46 |

**Table 7.** Comparisons of the ortholog groups identified by $3GC$ and $n3GC$ among *P. brasiliensis, C. albicans* and *S. cerevisiae*

| Genomes | $n3GC$ homologs | $3GC$ homologs | % | Identical groups | % | Coherent groups | % |
|---|---|---|---|---|---|---|---|
| Pb x Ca | 1011 | 482 | 209,75 | 266 | 26,31 | 278 | 27,50 |
| Ca x Sc | 4261 | 5024 | 84,81 | 3014 | 70,73 | 3088 | 72,47 |
| Sc x Pb | 154 | 186 | 82,80 | 50 | 32,47 | 52 | 33,77 |
| Pb x Ca x Sc | 3755 | 5028 | 74,68 | 2790 | 74,30 | 2883 | 76,78 |

**Table 8.** Comparisons of the ortholog groups identified by $3GC$ and $n3GC$ among *P. brasiliensis, C. neoformans* and *S. cerevisiae*

| Genomes | $n3GC$ homologs | $3GC$ homologs | % | Identical groups | % | Coherent groups | % |
|---|---|---|---|---|---|---|---|
| Pb x Cn | 1461 | 958 | 152,51 | 522 | 35,73 | 528 | 36,14 |
| Cn x Sc | 2276 | 3234 | 70,38 | 1382 | 60,72 | 1422 | 62,48 |
| Sc x Pb | 453 | 470 | 96,38 | 174 | 38,41 | 184 | 40,62 |
| Pb x Cn x Sc | 2957 | 4314 | 68,54 | 2175 | 73,55 | 2235 | 75,58 |

## 5   Conclusions

In this work we presented a new method, named $n3GC$, to infer biological functions by identifying homologous genes among three genomes, taking as input paralogous genes families of each one of the genomes. The EGG method [1] had been used to identify families of paralogous genes, and is now incorporated to the $n3GC$ method, but another method could be used, as well, to identify paralogs.

The results of $n3GC$ can be visualized using Venn diagrams, in which each set represents a genome. The number of homologs among the three genomes is shown in the intersection of the three sets, the number of homologs among two genomes is shown in the intersection of the corresponding two sets, and the number of genes exclusive to one genome is shown in a specific set region. Clicking in each region, details of the families and genes used to built the orthology relationships can be seen [16].

Besides, our method was used to identify potential pathogenic genes of the *P. brasiliensis* fungus, comparing it, three genomes at a time, with human pathogenic fungi - *C. albicans* and *C. neoformans*, and five human non-pathogenic fungi - *A. nidulans*, *F. graminearum*, *M. grisea*, *N. crassa* and *S. cerevisiae*. We observe that the *P. brasiliensis* dataset consists of assembled ESTs, instead of protein sequences as all the other organisms. Although we use this partial data to derive homology relationships, we are aware that accuracy may be diminished. The comparisons generated 10 Venn diagrams.

To validate the $n3GC$ method we compared some of its results with the results of Pfam [3], INPARANOID [13] and $3GC$ [15]. Comparing outputs of $n3GC$ applied to *P. brasiliensis*, *C. albicans* and *S. cerevisiae* genomes, and outputs of *P. brasiliensis*, *C. neoformans* and *S. cerevisiae* genomes with Pfam, we obtained approximately 95% of the relationships having 1 or 2 Pfam results, that is, genes identified as homologous by $n3GC$ have the same or a similar Pfam biological function, which suggests that our method is correctly indicating homology relationships. When comparing our results with INPARANOID and $3GC$ we obtained similar results, showing that the homology relationships were conserved.

Our method presented a good practical performance. The running time on a PC Pentium 4 1.5GHz with 512MB memory was 9.5 minutes on the average to each analysis of three genomes, with BLASTs already executed. Each genome had on the average 8540 genes.

For future work we suggest to include other algorithms to identify families of paralogs, such that we could offer on our site [16] a choice to the method to generate the paralog families. Also, some parameters could be passed to the program at execution, like values for the *e-value* and the coverage BLAST parameters. In the specific case of ESTs, these parameters could be optimized for accuracy. An environment with $n3GC$ execution directly from the site could be easily developed. We initially included it but as we did not implement an efficient system to allow multiple submissions, multiple requests used many computational resources, and led to a long time or even to an interruption of the services. So, a future task could be to manage multiple submissions. Particularly, we could use distributed computation to implement the presented methods. Another useful work is to improve the visualization of the results produced by the method, to facilitate analysis and interpretation of the obtained results is another useful work. Finally, our method could be used to identify other potential biological functions that could be investigated comparing genes of three organism.

# References

1. Almeida, N.F.: Tools for genome comparison (in Portuguese). PhD thesis, IC-UNICAMP (2002)
2. Altschul, S.F., Lipman, D.J., Madden, T.L., Miller, W., Schaffer, A.A., Zhang, J., Zhang, Z.: Gapped blast and PSI-BLAST: a new generation of protein database search programs. Nucleic Acid Research 25(17), 3389–3402 (1997)
3. Bateman, A., co-authors: The Pfam protein families database. Nucleic Acids Research 30(1), 276–280 (2002)
4. Birren, B.: Initiative Fungal Genome. A white paper for fungal comparative genomics. Whitehead Institute MIT Center for Genome (2003)
5. Braun, E.L., Halpern, A.L., Nelson, M.A., Natvig, D.O.: Large-scale comparison of fungal sequence information: mechanisms of innovation in Neurospora crassa and gene loss in Saccharomyces cerevisiae. Genome Research 10, 416–430 (2000)
6. Cannon, S.B., Young, N.B.: Orthoparamap: Distinguishing orthologs from paralogs by integrating comparative genome data and gene phylogenies. BioMed Central Bioinformatics 4, 35 (2003)
7. Delcher, A.L., Kasif, S., Fleischmann, R.D., White, O., Peterson, J., Salzberg, S.L.: Alignments of whole genome. Nucleic Acid Research 27(11), 2369–2376 (1999)
8. Felipe, M.S.S., co-authors: Transcriptional profiles of the human pathogenic fungus paracocidioides brasiliensis in mycelium and yeast cells. The Journal of Biological Chemistry 280(26), 24706–24714 (2005)
9. Kellis, M., Patterson, N., Birren, B., Berger, B., Lander, E.S.: Methods in comparative genomics: genome correspondence, gene identification and motif discovery. Bioinformatics 11(2-3), 319–355 (2004)
10. Lee, Y., Sultana, R., Pertea, G., Cho, J., Karamycheva, S., Tsai, J., Parvizi, B., Cheung, F., Antonescu, V., White, J., Holt, I., Liang, F., Quacjenbush, J.: Cross-referencing eukaryotic genomes: TIGR orthologous gene alignments (TOGA). Genome Research 12(3), 493–502 (2002)
11. Li, L., Stoeckert Jr, C.J., Roos, D.S.: OrthoMCL: Identification of ortholog groups for eukaryotic genomes. Genome Research 13(9), 2178–2189 (2003)
12. Liu, Y., Liu, X.S., Wei, L., Altman, R.B., Batxoglou, S.: Eukariotic regulatory element conservation analysis and identification using comparative genomics. Genome Research 14, 451–458 (2004)
13. Remm, M., Storm, C.E., Sonnhammer, E.L.: Automatic clustering of orthologs and in-paralogs from pairwise species comparisons. Journal of Molecular Biology 314(5), 1041–1052 (2001)
14. Tatusov, R.L., Natale, D.A., Garkavtsev, I.V., Tatusova, T.A., Shankavaram, U.T, Rao, B.S., Kiryutin, B., Galperin, M.Y., Fedorova, N.D., Koonin, E.V.: The COG database: new developments in phylogenic classification of proteins from complete genomes. Nucleic Acids Res 29, 22–28 (2001)
15. Telles, G.P., Brigido, M.M., Almeida, N.F., Viana, C.J.M., Anjos, D.A.S., Walter, M.E.M.T.: A method for comparing three genomes. In: Setubal, J.C., Verjovski-Almeida, S. (eds.) BSB 2005. LNCS (LNBI), vol. 3594, pp. 160–169. Springer, Heidelberg (2005)
16. website 3GC, http://egg.dct.ufms.br/n3gc/

# Validating Gene Clusterings by Selecting Informative Gene Ontology Terms with Mutual Information

Ivan G. Costa[1], Marcilio C. P. de Souto[2], and Alexander Schliep[1]

[1] Max-Planck-Institut für Molekulare Genetik,
Ihnestr., 63 D-14195 Berlin, Germany
{ivan.filho,alexander.schliep}@molgen.mpg.de
[2] Dep. de Informática e Matemática Aplicada - UFRN
Campus Universitário, 59072-970 Natal (RN), Brazil
marcilio@dimap.ufrn.br

**Abstract.** We propose a method for global validation of gene clusterings. The method selects a set of informative and non-redundant GO terms through an exploration of the Gene Ontology structure guided by mutual information. Our approach yields a global assessment of the clustering quality, and a higher level interpretation for the clusters, as it relates GO terms with specific clusters. We show that in two gene expression data sets our method offers an improvement over previous approaches.

**Keywords:** cluster validation, external index, gene ontology, mutual information.

## 1 Introduction

With the advent of DNA microarrays there has been a great deal of work on clustering methods for the analysis of data from large-scale gene expression experiments. The main idea behind these approaches is to find clusters of co-expressed genes, providing biologists with genes regulated in a similar manner [9]. While most of these approaches yielded useful analysis of gene expression data, the evaluation of the biological relevance of the clusters is still a difficult task. There is little guidance available for choosing a clustering method [8]. There is also no established framework for evaluation of gene clusterings resulting from these methods exists.

The biological interpretation of clusters has been addressed, for instance, by comparing the results with available functional genomics data, such as provided by the Gene Ontology (GO) project [2] (see Section 2.1 for more details). One common approach is to search for GO terms (functional annotations) that are significantly enriched within a cluster of genes [3,4]. Although this allows a biological interpretation of individual clusters of genes, it gives no global assessment on the "quality" of a gene clustering (or a set of clusters) returned by a clustering method.

M.-F. Sagot and M.E.M.T. Walter (Eds.): BSB 2007, LNBI 4643, pp. 81–92, 2007.

Recently, there have been proposals of global indices for the validation of gene clusterings [6,10]. However, in contrast to the approaches in [3,4], these validation methods provide no "biological" interpretation of their assessments. Furthermore, they do not take several important features of GO into account. For example, the GO structure (direct acyclic graph or DAG) presents a parent-child relation, which implies that a term inherits all annotations of its immediate descendent [1,11]. This makes the annotations of a GO term highly redundant with respect to terms "near" in the GO DAG. The use of redundant terms possibly introduces a bias in the global index, since contributions of GO terms that have many siblings will have a higher weight [10].

Motivated by the limitations presented above, we present a method that provides a global validation measure of gene clusterings. The method works by selecting a set of informative and non-redundant GO terms through an exploration of the Gene Ontology structure with the mutual information measure [7]. By informative, we mean terms that help to discriminate between clusters in a clustering. Additionally, by taking the parent-child relationship into account, our method detects a list of non-redundant GO terms within the informative ones. With this set of terms, we can calculate, as in [6,10], a global fitness measure of the clustering. Furthermore, our method relates a set of informative GO terms to a particular cluster, which provides a biological interpretation of the results.

## 1.1   Related Work

One of the first applications that used GO for evaluating groups of genes was the so called GO Term Enrichment (TE) analysis. By means of a statistical test, such as the Fisher exact test, one can estimate a $p$-value indicating whether a significant fraction of genes in a cluster is annotated with a specific GO term [3,4]. This approach has some limitations as it assumes independence between GO terms, and it suffers from the multiple testing problem [18]. More recent methods [1,11] take the dependencies of GO terms caused by the parent-child relations into account. In particular, the Parent-Term Enrichment method (PE) [11] assumes that whenever a particular term is enriched, so are its parents. Thus, it yields a more refined selection of GO terms.

All these methods have been shown useful and have found widespread use in the interpretation of individual clusters of genes. However, as previously mentioned, they do not produce a global assessment of how "biologically relevant" a given gene clustering is.

A global index for evaluating gene clusterings with GO was presented in [10]. This index, based on an approximation of mutual information, is able to discriminate between results of clustering methods from random cluster assignments. In [6], an external index was proposed for a similar task. However, neither of these two approaches account for any biological interpretation of the results. A further extension of [10] was presented in [17], where an informative set of GO terms is collected and the exact mutual information is computed. This method, however, has a high computational cost. It is exponential in the number of selected GO terms. Thus, in practice, only a small set of GO terms can be chosen.

Putting our approach into perspective, it combines the characteristics of "global indices", such as [10], with the interpretability of the "local" approaches such as [1,3,11]. Also, in contrast to [17], we constrain the selection of terms within the GO structure, yielding a more efficient computational procedure. This also makes the identification of redundant GO terms possible, which decreases bias of the global index towards GO terms having many siblings.

# 2  Method

## 2.1  Gene Ontology

The Gene Ontology (GO) project is a collaborative effort to address the need for consistent descriptions of gene products in different databases [2]. Three structured controlled vocabularies (ontologies) describe gene products in terms of their associated biological processes, cellular components and molecular functions in a species-independent manner—cellular component describes components in which genes are active (e.g., *rough endoplasmic reticulum*); molecular function contains concepts related to gene function (e.g., *catalytic activity*); and biological process describes the processes that a gene can take part of (e.g., *cellular physiological process*).

More formally, a given Gene Ontology (GO) is represented by a directed acyclic graph (DAG), in which each node $t_i$ in a set $T = \{t_1, ..., t_N\}$ represents a biological term (controlled vocabulary or GO term) and the edges stand for a set of relationships $\mathcal{R}$ among these terms. A relationship $R(t_i, t_j) \in \mathcal{R}$ means that term $t_i$ is a parent of term $t_j$. Such a relation is interpreted as $t_j$ is a subclass of $t_i$—i.e., $t_i$ is a more general concept than $t_j$. For instance, the biological term "*cell cycle*" is related to the more specific terms "*mitotic cell cycle*" and "*meiotic cell cycle*".

A set of genes $G = \{g_1, ..., g_M\}$ is related to a given GO by an annotation set $\mathcal{A}$, where $A(t_i, g_m) \in \mathcal{A}$ indicates that gene $g_m$ is annotated with term $t_i$. Genes often have multiple biological roles, so they are usually annotated with several GO terms. Furthermore, the parent-child relation of GO implies that genes annotated to a term are also annotated to all parents of this term. That is, for all $R(t_i, t_j) \in \mathcal{R}$, given a gene $g_m$, $A(t_j, g_m) \rightarrow A(t_i, g_m)$.

## 2.2  Selecting Informative GO Terms by Mutual Information Gain

In order to select a set of non-redundant and informative GO terms, we explore the DAG structure of GO and the parent-child relation. By informative terms we refer to terms that help to discriminate a cluster from others in a clustering. This can be measured with the mutual information, which is a general measure of dependence between two random variables [7]. In our case, the mutual information provides a systematic quantitative measure of the relationship between cluster membership and GO term membership of a set of genes. We call redundant terms the ones that annotate a similar set of genes. Recall the parent-child relation $R(t_i, t_j)$, as $t_i$ also annotates all terms $t_j$ annotates, we expect that $t_i$ is informative whenever $t_j$ is.

Our selection procedure—called MutSel—works bottom-up as follows. For a given GO, a set of genes $G$ and its respective annotation set $\mathcal{A}$, we start with a candidate collection of terms $S$ with unitary sets, each one containing a leaf node (a node of the DAG without descendants). Such a collection corresponds to the most specific annotations present in GO for genes in $G$. From these we calculate the gain in mutual information, with respect to the cluster membership, when joining each set $\mathbf{s}_i \in S$ either with other adjacent (or neighboring) set or with parent terms not included in the candidate sets $S$.

The set of adjacency relations, $\mathcal{D}$, is defined by the parent-child relation, where sets $\mathbf{s}_p$ and $\mathbf{s}_q$ are adjacent, $D(\mathbf{s}_p, \mathbf{s}_q)$, if and only if there exists terms $t_i \in \mathbf{s}_p$ and $t_j \in \mathbf{s}_q$, such that $R(t_i, t_j) \in \mathcal{R}$ or $R(t_j, t_i) \in \mathcal{R}$. At each step, we select the pair of adjacent sets that yields the higher non-negative mutual information gain, joining them in a new set of terms. This step is equivalent to looking for more general terms in the GO DAG, which are more informative to the clustering results. We repeat this step until no mutual information gain is possible.

More formally, let $X^p$ be a discrete random variable with alphabet $\mathcal{X} = \{0, 1\}$ representing the annotation of $\mathbf{s}_p$, where an observation $x$ takes the value 1 if a term in $\mathbf{s}_p$ annotates it, or zero otherwise. Respectively, the random variable $Y$ with alphabet $\mathcal{Y} = \{1, ..., K\}$ represents the cluster assignment, where a observation $y$ takes value $k$ if it belongs to cluster $k$. The mutual information gain, $\mathrm{MIG}(X^p, X^q|Y)$, of joining two adjacent sets $\mathbf{s}_p$ and $\mathbf{s}_q$ in the context of cluster membership $Y$ is defined as

$$\mathrm{MIG}(X^p, X^q|Y) = \mathrm{MI}(X^p \vee X^q, Y) - \mathrm{MI}(X^p, Y) - \mathrm{MI}(X^q, Y), \qquad (1)$$

where MI denotes the mutual information, and $X^p \vee X^q$ the variable resulting in the union of sets $s_p$ and $s_q$. The mutual information, MI, is defined as,

$$\mathrm{MI}(X^i, Y) = \sum_{x \in \mathcal{X}} \sum_{y \in \mathcal{Y}} \mathbf{P}[X^i = x, Y = y] \log \left( \frac{\mathbf{P}[X^i = x, Y = y]}{\mathbf{P}[X^i = x]\mathbf{P}[Y = y]} \right), \qquad (2)$$

$\mathrm{MI}(X^i, Y) \geq 0$, with equality only if both variables $X^i$ and $Y$ are independent.

For a given set of genes $G$, we have a set of observations $\{x_1^i, ..., x_M^i\}$, where $x_m^i = 1$ if $t^i$ annotates gene $m$, 0 otherwise. Respectively, we have a set of observations $\{y_1, ..., y_M\}$, where $y_m = k$ denotes that gene $m$ belongs to cluster $k$. From these observations, we can obtain the following estimates for computing $\mathrm{MI}(X^i, Y)$,

$$\mathbf{P}[X^i = j, Y = k|G] = \frac{1}{M} \sum_{m=1}^{M} 1\{x_m^i = j\} 1\{y_m = k\}, \qquad (3)$$

$$\mathbf{P}[Y = k|G] = \frac{1}{M} \sum_{m=1}^{M} 1\{y_m = k\} \qquad (4)$$

where 1 is a indicator function, $j \in \mathcal{X}$ and $k \in \mathcal{Y}$.

Figure 1 illustrates our method. On the left (Figure 1 (a)), we depict a simple example of a DAG with 7 terms. In Figure 1(b), we display a table, where the rows corresponds to the random variables $X^i$ and the columns the genes from set $G$. An one in position $(i,j)$ indicates that gene $j$ is annotated with term $i$. The last line, $Y$, indicates the assignment of genes to one of the two the clusters considered. At each node of the DAG in Figure 1(a), we display the cluster counts and the mutual information of the respective term. For example, in Term7, "1/3" means that this term annotates one gene from cluster 1 and three genes from cluster 2. The value 0.258 corresponds to the mutual information. Terms with good discriminative power in relation to $Y$ display a higher MI (e.g., Term2 and Term7) than non-discriminative terms (e.g Term1 and Term4).

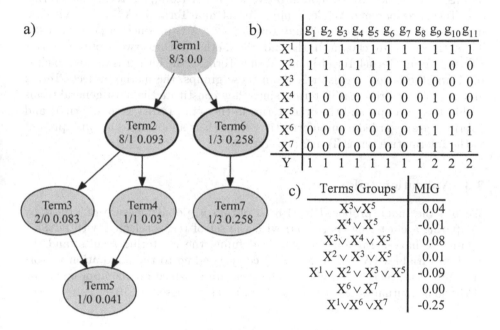

**Fig. 1.** We depict on the left (a) an example of a simple DAG, on the top right (b) a table describing the terms annotations to a set of 11 genes and on the bottom right (c) a list of candidate join operations and the respective MIG

Starting with a collection $S = \{s_1, ..., s_P\}$ such that $s_p = \{t_l\}$ where $t_l$ is a leaf from GO DAG, and $\mathcal{D}$ is the adjacency list, the algorithm works as follows:

1. while $\max_{D(s_i,s_j)\in\mathcal{D}} MIG(X^i, X^j|Y) \geq 0$ do
2.    $D(s_p, s_q) = \arg\max_{D(s_i,s_j)\in\mathcal{D}} MIG(X^i, X^j|Y)$
3.    join$(s_p, s_q)$
4.    update$(\mathcal{D})$

The algorithm returns a collection $S$ of groups of GO terms. Given that we join only parent terms, each of these groups constitutes a sub-DAG from GO. From these, we can select the most general terms, or the terms without any parent node within a group $s_p$ as the representative term(s) of $s_p$. All other terms in the group can be considered as redundant, since they will carry the same or less information than the representative terms. Furthermore, we can also relate a given group of terms $s_p$ with a cluster in $k'$

$$k' = \arg\max_{k \in \mathcal{Y}} \text{MI}(X^p, Y = k). \tag{5}$$

Figure 1(c) illustrates a simple example of the method. There, we display the MIG from joining candidate sets of terms. The selection method starts with the leaf nodes Term5 and Term7. It then looks for neighboring terms, whose unions with the leaves has non-negative MIG. For example, Term5 has Term3 and Term4 as parents. While joining Term4 and Term 5 ($X^4 \vee X^5$) yields a negative MIG, merging Term3 and Term5 ($X^3 \vee X^5$) produces a positive MIG. Thus, the latter are chosen. In the end, the method returns two groups of terms {Term2,Term3,Term4,Term5} and {Term6,Term7}: the former is related to cluster 1 and the latter to cluster 2. From these groups, the method selects Term2 and Term6 as representative terms, since they constitute the most general terms within these groups; and the other terms in the sets {Term3,Term4,Term5} and {Term7} are regarded as uninformative, since their annotations are also present in the informative terms Term2 and Term6.

## 2.3   Validation Index

We use the index proposed in [10] to obtain a global measure of fitness by comparing a clustering (partition) with the set of terms selected with MutSel. Again, we have a random variable $Y$ defining the clustering results, and the random variables $\{X^1, ..., X^p, ..., X^P\}$ corresponding to the annotation vectors of group of terms selected above. The measure in based on the approximation of the joint mutual information $\text{MI}^{app}(X, Y)$ as proposed in [10],

$$\text{MI}^{app}(X, Y) = \sum_{p=1}^{P} \text{MI}(X^p, Y). \tag{6}$$

As discussed in [17], this approximation assumes independence between variables from $X$, which does not hold for most selections of GO terms, given the high dependency between GO term annotations. An alternative to improve the approximation of Eq. 6 is to select a set of terms with low annotation redundancy. To tackle this problem, [10] introduces a parameter $U$, also based in the mutual information, which excludes redundant terms from the computation. The smaller the value of $U$ is, the less redundancy will be allowed in the set of terms $X$ used for computing Eq. 6. Note that MutSel joins terms displaying dependencies caused by the parent-child property of GO annotations in a principled fashion, automatically excluding redundant terms and requiring no extra parameter.

To quantify deviation from randomness, we compute a $z$-score by repeating the MutSel procedure with random cluster assignments as performed in [10]. The random clusterings are draw with the same cluster size distribution as the evaluated clustering. More formally, from a given real clustering $Y$, its selection of GO terms $X$, a random clustering $Y^r$, its selection of GO terms $X^r$, then we have,

$$z\mathrm{MI}^{app} = \frac{\mathrm{MI}^{app}(X,Y) - \mu^r}{\sigma^r}. \tag{7}$$

where $\mu^r = \mathrm{Mean}(\mathrm{MI}^{app}(X^r, Y^r))$ is the mutual information mean for $L$ random clusterings and $\sigma^r = \mathrm{Var}(\mathrm{MI}^{app}(X^r, Y^r))^{1/2}$ is the standard deviation of the mutual information from $L$ random clusterings. Hereafter, we refer to $z^{MI}$ as the index proposed [10], and $z^{MutSel}$ as the index from Eq. 7 after selection of GO terms by MutSel.

# 3  Experiments

We evaluate our method on two typical scenarios of gene expression data analysis. First, we inspect the selection of GO terms in a differential gene expression analysis, where a group of induced and a group of repressed genes after treatment of yeast were identified [13]. This data, where two clusters of genes are given beforehand and no clustering analysis is needed, allow us to evaluate the "biological relevance" of the selection of GO terms, since the biological processes behind these two clusters are well characterized. In the second experiment, we perform a small scale comparison of clustering methods on a yeast cell cycle data set. This data set has been manually labeled [5], allowing us to compare our index and the prior approach [10] to the expert manual annotation.

## 3.1  Yeast Treatment (YT)

Gene expression of yeast was measured at particular time points after the treatment with sulfometuron methyl (SM) [13]. We use a group of 241 induced genes and a group of 121 repressed genes 4h after treatment with $5\mu g/ml$ of SM. This clustering gives a simple scenario to evaluate our method, since the biological processes behind theses two clusters are well characterized [13].

## 3.2  Yeast Cell Cycle (YCC5)

This dataset represents the expression levels of over 6,000 genes during two cell cycles from Yeast measured in 17 time points [5]. We used a subset YCC5, of 384 genes visually identified to peak at five distinct time points [5], each representing a distinct phase of cell cycle (Early G1, Late G1, S, G2 and M). Hereafter, this subset will be referred to as YCC5. The expression values of each gene were standardized, which can enhance the performance of model-based clustering methods, when the original data consists of intensity levels.

In relation to the clustering methods, we performed analysis with hierarchical clustering (`Hier`) [9], $k$-means [15], mixture of multivariate Gaussians with diagonal covariance matrix (`MixGaus`) [14] and mixtures of Hidden Markov models (`MixHMM`) [16]. We set the number of clusters to be equal to 5 in all methods (as this is the number of classes in the manual annotation). For $k$-means, `MixGauss` and `MixHMM`, we initialize models randomly, perform clustering 15 times, and selected the solution with minimal error criteria (see [16] for details). For $k$-means and hierarchical clustering, we used Pearson correlation as a similarity measure.

## 4  Results

### 4.1  GO Term Selection

In order to evaluate our method with respect to the selection of "biologically relevant" GO terms, we use the set of repressed and induced genes from the study on response of yeast to a inhibitor of amino acid synthesis [13] introduced in Section 3.1. Table 1 depicts the top five informative GO terms, from the Biological Process GO, for the induced genes (first five rows), as well as for the repressed ones (last five rows). The columns represent the GO term id, the GO term name, the counts of induced genes, the counts of repressed genes, and the mutual information.

As highlighted in [13], induced genes were mainly related to molecule transport, amino acid biosynthesis and nitrogen metabolism. Indeed, all terms from Table 1, with exception of *"vitamin biosynthetic process"*, are directly related to these processes. Among the repressed genes, the study detected genes related to carbohydrate and lipid biosynthesis, translation, cell cycle and ribosome. All terms listed in Table 1 bottom are either directly related or more general terms describing these processes.

**Table 1.** Top five informative GO terms, from the Biological Process GO, for induced (top) and repressed (bottom) genes

| Term ID | Term Name | #I | #R | MI |
|---------|-----------|----|----|----|
| GO:0006807 | nitrogen compound metabolic process | 68 | 14 | 0.022 |
| GO:0009110 | vitamin biosynthetic process | 14 | 0 | 0.022 |
| GO:0006519 | amino acid and derivative metabolic process | 60 | 13 | 0.018 |
| GO:0016769 | transferase activity, transferring nitrogenous groups | 11 | 0 | 0.017 |
| GO:0009059 | macromolecule biosynthetic process | 16 | 36 | 0.051 |
| GO:0051301 | cell division | 3 | 10 | 0.019 |
| GO:0008610 | lipid biosynthetic process | 4 | 11 | 0.018 |
| GO:0022613 | ribonucleoprotein complex biogenesis and assembly | 2 | 7 | 0.014 |
| GO:0044265 | cellular macromolecule catabolic process | 9 | 14 | 0.013 |

We also compare, in the context of `YT`, the GO terms selected with `MutSel` with the ones obtained with well-known methods, such as the Term Enrichment (`TE`) [3] and the Parent-Term enrichment (`PE`) [11]. Table 2 summarizes this comparison. Its rows correspond to, respectively, the set of induced and repressed genes in the dataset YT. The columns of the first part correspond to, respectively, the number of terms selected with PE ($p$-value lower then 0.05), the number of terms selected with MutSel, and the intersection of both sets. Likewise, in following columns, we present the number of terms selected with TE ($p$-value lower then 0.05), the number of all terms (informative and redundant) selected with MutSel (we refer to this set as MutSelAll), and the intersection of both sets.

Analyzing the results presented in Table 2, the informative terms selected by our method are mainly a smaller subset of genes enriched in PE; 85% of terms related to the cluster of induced genes and 81% of terms related to the cluster of repressed genes detected by MutSel are also selected in PE. Likewise, the result obtained with TE, which does not filter redundant terms, is comparable to the set of all terms (informative and redundant) selected by MutSel. Again, the terms indicated by the MutSelAll was a small subset of PE; 84% for the cluster of induced genes and 100% for the cluster of repressed genes.

To further investigate the distinction between these methods, we measure the redundancy of annotation of GO terms. For two GO terms, redundancy can be measure by computing their mutual information (MI): redundant terms have higher mutual information values. More precisely, we compute the mutual information between all pairs of GO terms from a given set, select the maximum MI for each term and average the values. For the cluster of induced genes, terms obtained with MutSel, MutSelAll, PE, and TE had a MI mean of, respectively, 0.154, 0.219, 0.271, and 0.275. For the set of repressed genes, these values were, respectively, 0.097, 0.168, 0.198, and 0.185. In both cases, the methodologies taking the parent-child property into account displayed lower MI than their counterparts. In general, MutSel presented lower MI values, which demonstrates its ability to select a set of non-redundant terms.

**Table 2.** Comparison of the number of GO terms selected with MutSel, MutSelAll, TE and PE in the analysis of dataset YT

| | PE | MutSel | ∩ | TE | MutSelAll | ∩ |
|---|---|---|---|---|---|---|
| Induced | 41 | 13 | 11 | 79 | 39 | 33 |
| Repressed | 79 | 22 | 18 | 159 | 80 | 80 |

## 4.2 Comparison of Clustering Methods

We display in Table 3, for dataset YCC5, the rankings of the results from the four clustering methods, according to the different indices. More precisely, we list the rank of the methods according to $z^{MI}$ for five choices of $U$ and $z^{MutSel}$. After each method name we display the mean values for 10 replications of the $z$ score. The last line corresponds to the corrected Rand (CR) [12] of comparing the clustering assignment with the manual labeling. We used the original implementation to obtain values from $z^{MI}$ (available at http://llama.med.harvard.edu/cgi/cgi/ClusterJudge/cluster_judge.pl).

**Table 3.** We list the rank of the methods given by the indices $z^{MI}$ for several choices of $U$, $z^{MutSel}$, and $CR$ comparing the clustering assignment with the manual labeling

| Indices | Rank 1 | Rank 2 | Rank 3 | Rank 4 |
|---|---|---|---|---|
| $z^{MI}$ $U = 0.8$ | $k$-means (3.26) | MixHMM (2.87) | Hier. (2.86) | MixGaus (2.32) |
| $z^{MI}$ $U = 0.4$ | $k$-means (3.74) | Hier. (2.91) | MixHMM (2.87) | MixGaus (2.26) |
| $z^{MI}$ $U = 0.2$ | $k$-means (1.41) | MixHMM (0.27) | MixGaus (0.06) | Hier. (-0.17) |
| $z^{MI}$ $U = 0.1$ | $k$-means (0.86) | MixGaus (0.37) | MixHMM (0.36) | Hier. (-0.1) |
| $z^{MI}$ $U = 0.01$ | $k$-means (1.4) | MixGaus (0.83) | Hier. (0.64) | MixHMM (-0.1) |
| $z^{MutSel}$ | $k$-means (1115.3) | MixGaus (1034.0) | MixHMM (791.9) | Hier. (616.3) |
| CR | $k$-means (0.5) | Hier. (0.46) | MixGaus (0.43) | MixHMM (0.39) |

In general, $k$-means was ranked as the first one by all indices. In contrast, all others ranking positions differed from index to index. One important result that can be observed in this table is the impact of parameter $U$, the uncertainty index, in the values obtained by $z^{MI}$ [10] and on the resulting rankings. For instance, for higher $U$ values, where some redundancy in annotation is allowed, hierarchical clustering was ranked second; for more stringent values of $U$ (i.e., 0.1 and 0.2), the result of this algorithm presented a negative $z^{MI}$ score, which indicates results obtained by chance. These results contradict the claims in [10], where the authors state that the parameter $U$ had small influence on the rankings of methods. In comparison to $z^{MI}$, $z^{MutSel}$ yielded higher z-scores. This is explained by the fact that for random clusterings, MutSel makes very few merging operations. In this situation, the resulting selection of terms is mainly composed of leaf terms with few annotated genes. These terms have also very low information regarding $Y$. In other words, MutSel can easily discriminate clusterings from random generated ones.

No index was able to recover the ranking given by $CR$. Although we cannot take the annotation used to calculate the $CR$ as the actual and only "ground truth" for dataset YCC5, since it was made via visualization of profiles, such an annotation still provides a basis for comparing the clusterings. With regard to $z^{MutSel}$, the difference was mainly in the ranking of the hierarchical clustering. An inspection of the contingency table, cluster against annotation labels, shows that hierarchical clustering placed genes that correspond to two different classes of the manual annotation (phases S and G2) into a single cluster, and had a small cluster with 10 genes from all distinct classes. On the other hand, the other clustering solutions had no such small cluster. This indicates that $z^{MutSel}$ penalize this merge of groups S and G2 more strongly than $CR$. On the other hand, $z^{MI}$ did not yield a definitive solution, while its rankings vary from values of $U$. Furthermore, for lower $U$s, the value of the index for the hierarchical clustering are negative, which indicates that its results are comparable with a random solution. This strongly contradicts the $CR$ values derived from the manual annotation. As manual annotation is usually not provided in the majority of gene expression data sets, $z^{MutSel}$ represents a better alternative to $z^{MI}$, since it requires no extra parameters, while it selects the set of most informative and non-redundant terms.

# 5   Conclusion

In this paper, we present the MutSel method for computing a global validity measure of a clustering of genes. The main advantage of this method is a selection of relevant and non-redundant terms in relation to the evaluated clustering. In order to do so, we use of a characteristic intrinsic to Gene Ontology (GO), the parent-child relation, which makes annotations of GO terms highly redundant. The set of informative and non-redundant GO terms resulted from the application of MutSel yields not only a global index of "biological validity" of the clustering, but it also relates GO terms to clusters yielding a "biological interpretation" of individual clusters.

A comparison of MutSel to established methods for providing interpretation of a cluster of genes, such as Term Enrichment analysis and Parent-Term Enrichment analysis, showed that MutSel mainly selects a set of GO terms also found to be relevant by these methods. Furthermore, the set of selected terms has a lower degree of annotation redundancy.

In relation to a global evaluation index for clusterings, we show that the selection of terms from MutSel improves the mutual information-based measure proposed in [10]. Our experimental results show that the choice of parameters of the original index [10] has a great impact on the resulting rankings of clustering methods. Thus, MutSel represents an improvement to the original proposal, as it requires no parameter settings, while its results are consistent with manual annotation of genes in a benchmark data set. As an extension of this work, we plan to accomplish a large scale evaluation, including more clustering methods and gene expression data sets.

**Acknowledgments.** The first author would like to acknowledge funding from the DAAD/CNPq (Brazil).

# References

1. Alexa, A., Rahnenfuhrer, J., Lengauer, T.: Improved scoring of functional groups from gene expression data by decorrelating GO graph structure. Bioinformatics 22(13), 1600–1607 (2006)
2. Ashburner, M.: Gene ontology: tool for the unification of biology. Nat Genet 25(1), 25–29 (2000)
3. Beissbarth, T., Speed, T.P.: GOstat: find statistically overrepresented Gene Ontologies within a group of genes. Bioinformatics 20(9), 1464–1465 (2004)
4. Boyle, E.I., Weng, S., Gollub, J., Jin, H., Botstein, D., Cherry, J.M., Sherlock, G.: GO:TermFinder–open source software for accessing Gene Ontology information and finding significantly enriched Gene Ontology terms associated with a list of genes. Bioinformatics 20(18), 3710–3715 (2004)
5. Cho, R.J., Campbell, M.J., Winzeler, E.A., Steinmetz, L., Conway, A., Wodicka, L., Wolfsberg, T.G., Gabrielian, A.E., Landsman, D., Lockhart, D.J., Davis, R.W.: A genome-wide transcriptional analysis of the mitotic cell cycle. Mol Cell 2(1), 65–73 (1998)

6. Costa, I.G., Schliep, A.: On external indices for mixtures: validating mixtures of genes. In: Spiliopoulou, M., Kruse, R., Borgelt, C., Nurnberger, A., Gaul, W. (eds.) From Data and Information Analysis to Knowledge Engineering, pp. 662–669. Springer, Heidelberg (2005)
7. Cover, T.M., Thomas, J.A.: Elements of Information Theory. Wiley - Interscience, Chichester (1991)
8. D'haeseleer, P.: How does gene expression clustering work? Nat Biothech 24(12), 1499–1501 (2005)
9. Eisen, M.B., Spellman, P.T., Brown, P.O., Botstein, D.: Cluster analysis and display of genome-wide expression patterns. PNAS 95(25), 14863–14868 (1998)
10. Gibbons, F.D., Roth, F.P.: Judging the Quality of Gene Expression-Based Clustering Methods Using Gene Annotation. Genome Res. 12(10), 1574–1581 (2002)
11. Grossmann, S., Bauer, S., Robinson, P.N., Vingron, M.: An improved statistic for detecting over-represented gene ontology annotations in gene sets. In: Apostolico, A., Guerra, C., Istrail, S., Pevzner, P., Waterman, M. (eds.) RECOMB 2006. LNCS (LNBI), vol. 3909, pp. 85–98. Springer, Heidelberg (2006)
12. Hubbert, L.J., Arabie, P.: Comparing partitions. Journal of Classification 2, 63–76 (1985)
13. Jia, M.H., LaRossa, R.A., Lee, J.-M., Rafalski, A., DeRose, E., Gonye, G., Xue, Z.: Global expression profiling of yeast treated with an inhibitor of amino acid biosynthesis, sulfometuron methyl. Physiol. Genomics 3(2), 83–92 (2000)
14. McLachlan, G., Peel, D.: Finite Mixture Models. Wiley Series in Probability and Statistics. Wiley, New York (2000)
15. McQueen, J.: Some methods of classification and analysis of multivariate observations. In: 5th Berkeley Symposium in Mathematics, Statistics and Probability, pp. 281–297 (1967)
16. Schliep, A., Costa, I.G., Steinhoff, C., Schonhuth, A.: Analyzing gene expression time-courses. IEEE/ACM Trans. Comput. Biol. Bioinformatics 2(3), 179–193 (2005)
17. Steuer, R., Humburg, P., Selbig, J.: Validation and functional annotation of expression-based clusters based on gene ontology. BMC Bioinformatics 7(1), 380 (2006)
18. Westfall, P., Young, S.: Resampling-Based Multiple Testing: Examples and Methods for p-Value Adjustment. Wiley-Interscience, Chichester (1993)

# An Optimized Distance Function for Comparison of Protein Binding Sites

Gábor Iván

Department of Computer Science, Eötvös University,
Pázmány Péter stny. 1/C, H-1117 Budapest, Hungary
hugeaux@cs.elte.hu.

**Abstract.** An important field of application of string processing algorithms is the comparison of protein or nucleotide sequences. In this paper we present an algorithm capable of determining the dissimilarity (*distance*) of protein sequences originating from protein binding sites found in the RS-PDB database that is a repaired and cleaned version of the publicly available Protein Data Bank (PDB). The special way of construction of these protein sequences enabled us to optimize the algorithm, achieving runtimes several times faster than the unoptimized approach. One example the algorithm proposed in this paper can be useful for is searching conserved sequences in protein chains.

## 1 Introduction

String comparison algorithms have a wide range of applications. Among the most important ones is the comparison of protein and nucleotide sequences. These algorithms work by aligning the two strings under comparison, while maximizing or minimizing a cost function. Cost functions are usually defined by substitution matrices that assign costs to character-pairs and a function describing additional costs when inserting gaps. Some algorithms are guaranteed to find the optimum of the cost function; some heuristic algorithms find a value very close to the optimum with a high probability. A brief summary of substitution matrices, gap penalty types and sequence comparison algorithms is given below.

### 1.1 Scoring Matrices

Scoring matrices assign a score to all possible amino acid-pairs. The score can, for instance, correlate with the probability that a particular amino acid-type transforms into another amino acid-type due to mutation (PAM (Point Accepted Mutation) matrix). Another, frequently used scoring matrix type is BLOSUM (BLOcks SUbstitution Matrix) [3]. Scoring matrices are usually square, and – in the case of amino acids – have at least 20 rows/columns, as in most cases they contain scores for the 20 standard amino acids.

### 1.2 Gap Penalty Types

There are several types of costs that can be assigned to insertion of gaps into an amino acid sequence. A gap with a length of $k$ characters can be penalized as follows:

M.-F. Sagot and M.E.M.T. Walter (Eds.): BSB 2007, LNBI 4643, pp. 93–100, 2007.

- Linear gap penalty: $g(k) = a \cdot k$;
- Affine gap penalty: $g(k) = b + a \cdot k$;
- Other, but monotone (e.g., logarithmic) function: $g(k) = b + a \cdot \log k$. The application of monotone functions has considerable advantages, as it was shown in [4].

## 1.3   Algorithms Finding the Optimal Alignment Score

After defining gap costs and replacement scores for amino acids, sequence alignment algorithms are required to align two amino acid sequences so that an optimal alignment score can be achieved. Smith-Waterman [6] and Needleman-Wunsch [5] algorithms are an example; the former implements local, the latter global alignment. Given a certain scoring scheme, these algorithms always find the optimum score, but they are also somewhat time-consuming. Note that in both of the mentioned algorithms, *optimum* score means *maximum*. This is important to emphasize, because later a function capable of measuring the distance of sequences will be explained; in that case, the aim will be to find the *minimum* distance.

The above mentioned algorithms are usually implemented as DPAs (Dynamic Programming Algorithms).

## 1.4   Heuristic Algorithms

Heuristic algorithms do not necessarily find the optimal alignment score. Instead, they try to keep the probability of missing a high-scoring alignment acceptably low, while being at least one order of magnitude faster than the algorithms guaranteed to find the optimal score mentioned before. BLAST (Basic Local Alignment Search Tool) [9] is a basic example of such heuristic algorithm that is used to compare a single amino acid sequence to a large sequence database. BLAST assumes that a high-scoring local alignment contains two short (e.g., with a length of three amino acids) very similar subsequences with a high probability.

There were several improvement attempts for the original BLAST algorithm; examples are Gapped BLAST and PSI-BLAST [7].

## 1.5   Aim

Our purpose was to implement a function that measures the distance (dissimilarity) of two amino acid sequences extracted from ligand binding sites. The sequences were constructed in a way described in Section 2.2. This construction of sequences enabled us to optimize the distance function, drastically improving (reducing) required runtime.

A distance function on a given set (in our particular case, this set contains protein chains) has to satisfy the following conditions:

1. $d(x, y) \geq 0$ (non-negativity)
2. $d(x, y) = 0$ if and only if $x = y$ (identity of indiscernibles)
3. $d(x, y) = d(y, x)$ (symmetry)
4. $d(x, z) \leq d(x, y) + d(y, z)$ (subadditivity/triangle inequality)

# 2  The Algorithm

As our algorithm works specifically on amino acid sequences originating from binding sites found in the RS-PDB database [1], it is necessary to define how these sequences were extracted from the database.

## 2.1  The Definition of Binding Sites

Binding sites are defined as a set of atom-pairs; one atom belongs to some protein (described by amino acid sequence), and one atom to some ligand in each. Distance of the two atoms - with some tolerance - has to be equal to the sum of Van der Waals radii calculated for the atoms (depending on their type). By using this definition for binding sites, all amino acids from a given amino acid sequence that have at least one atom contained in an atom pair-set (describing some binding site) can be marked. These amino acids are called *binding amino acids*. Binding sites were extracted from the RS-PDB database described in [1], [2].

A somewhat "typical" binding site found in PDB entry 1QD6 can be seen on Figure 1; the protein part is an enzyme from *Escherichia coli*.

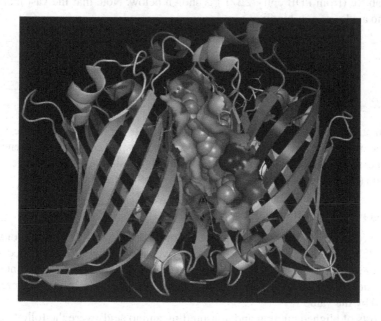

**Fig. 1.** A binding site visualized from PDB entry 1QD6

## 2.2  Amino Acid Sequence Representation

By *amino acid sequence* we mean sequences consisting of amino acids connected by peptide bonds that are of maximal length (i.e. they cannot be continued with further amino acids on either end). It has to be noted that multiple amino acid sequences might occur in the immediate vicinity of a single binding site. In this paper, however, our goal is to deal with single amino acid sequences, defining their distance and proposing an algorithm capable of calculating this measure.

A *binding amino acid* is an amino acid if it has at least one common binding atom-pair with a ligand. A *binding amino acid sequence* is an amino acid sequence that contains at least one *binding amino acid*.

Binding amino acid sequences are extracted from protein binding sites as follows:

A string is assigned to each amino acid sequence binding to some ligand, the characters of which correspond to amino acids of the given sequence. In this string, amino acids participating in the bond are indicated by their one-character code; non-binding amino acids are indicated by ' − ' characters. As our purpose was to deal with only binding sections, we omitted pre- and postfixes of amino acid sequences purely consisting of non-binding amino acids (or, according to our current notation, ' − ' characters). Hence all the strings constructed this way start and end with a binding amino acid.

Binding amino acids can be considered to be given weights to emphasize their significance. Non-binding amino acids are aggregated into one symbol (here, the ' − ' character).

**Example.** A *binding amino acid sequence* constructed and transformed the way described above (from PDB entry 2BZ6) is shown below. Note that the vast majority of the amino acids are non-binding:

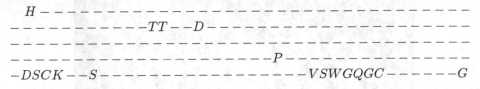

The motivation (or biological justification) for this representation is the usage of binding amino acids for marking start/endpoints of possible conserved sequences (*motifs*). A possible application might be the clustering of similar (low-distance) binding sites while (at the same time) finding evolutionally related sequences.

### 2.3 Algorithm Definition

For measuring the distances of binding sections of amino acid sequences constructed the way described above, we used a modified version of the algorithm used for calculating *Levenshtein-distance*. The modifications involved assigning different costs to gaps depending on where they are inserted, while amino acid mismatches were simply penalized by the value 1.

The costs of aligned binding and non-binding amino acids were the following:

- The cost of two aligned, different amino acids is 1.
- The cost of aligned, matching amino acids is zero.

Gaps were penalized as follows:

- Insertion of a gap with a length of one unit (one amino acid) costs *gp* (*gap penalty*), if the gap is aligned with a non-binding amino acid in the other sequence. If a gap is aligned with a binding amino acid, its cost is 1.

- Insertion of gaps at the end of sequences is only penalized if they are aligned with binding amino acids. Gaps inserted at either end of a sequence have a zero cost, if they are aligned with non-binding amino acids. The linear gap penalty used in our algorithm will be furthermore referred to as $GP$.

According to the costs defined above, the rules applied during filling the first and last rows/columns of the matrix used by the DPA have also been changed as well.

A $D$ matrix is constructed consisting of $L_1 + 1$ columns and $L_2 + 1$ rows, where $L_1$ and $L_2$ denotes the length of the two sequences ($seq_1$, $seq_2$) in comparison.

$D[i, j]$ means the minimum cost of transforming the $i$-length prefix of $seq_2$ to the $j$-length prefix of $seq_1$.

Initialization of the $D$ matrix is performed according to the following rules:

$$D[0,j] = \begin{cases} D[0, j-1] + 1 & \text{if } seq_1[j] \neq \text{'-'}, \text{otherwise} \\ D[0, j-1] \end{cases}$$

$$D[i,0] = \begin{cases} D[i-1, 0] + 1 & \text{if } seq_2[i] \neq \text{'-'}, \text{otherwise} \\ D[i-1, 0] \end{cases}$$

Rules that determine filling of matrix cells:

$$x = \begin{cases} D[i-1, j] & \text{if } seq_2[i] = \text{'-' and } j = L_1, \text{otherwise} \\ D[i-1, j] + GP & \text{if } seq_2[i] = \text{'-', otherwise} \\ D[i-1, j] + 1. \end{cases}$$

$$y = \begin{cases} D[i, j-1] & \text{if } seq_1[j] = \text{'-' and } i = L_2, \text{otherwise} \\ D[i, j-1] + GP & \text{if } seq_1[j] = \text{'-', otherwise} \\ D[i, j-1] + 1. \end{cases}$$

$$z = \begin{cases} D[i-1, j-1] & \text{if } seq_1[j] = seq_2[i], \text{otherwise} \\ D[i-1, j-1] + 1 \end{cases}$$

$D[i, j] = min(x, y, z)$.

For easier understanding of the motivation behind the rules above, we can consider all sequences to be concatenated with a sequence consisting of an infinitesimal number of non-binding amino acids on both ends. Hence, a zero-length amino acid sequence can be considered an infinitesimally long sequence consisting of non-binding amino acids. This makes sense because we do not want to penalize gaps if they are inserted at either ends of amino acid sequences, and a possible way of achieving this is inserting an infinite number of imaginary non-binding amino acids on both ends of sequences.

It might be interesting to notice that the quantity determined by the above algorithm fulfills the criteria for being a metric – this is why we take the courage to call it "distance". The proof of the metric property is very similar to that of *Levenshtein distance*.

## 3  Optimization

The optimization of the algorithm is based on the observation that sequences constructed the way described in Section 2.2 usually contain fairly long subsequences consisting of non-binding amino acids. This equals to long subsequences consisting of '-' characters.

It can be proven that, using the above defined costs for amino acid mismatches and gaps, determining the value of matrix cells (in the $D$ matrix that is used by DPA for distance calculation) is rather easy at cells that are located at the intersection of two subsequences consisting of non-binding amino acids: in this case, the only choice is to step diagonally.

This can be proven if we consider that at the intersection of two non-binding amino acids, the DPA algorithm chooses the *minimum* from the following values when filling $D[i,j]$:

$$\begin{cases} D[i-1,j-1] & \text{(diagonal step)} \\ D[i,j-1] + GP \\ D[i-1,j] + GP \end{cases}$$

If either $D[i,j-1]$ or $D[i-1,j]$ contained a significantly lower value than $D[i-1,j-1]$, the DPA would of course not choose the latter for $D[i,j]$. In the following section we prove that (using the costs defined in Section 2.3) neither $D[i,j-1]$ nor $D[i-1,j]$ can contain a value that is smaller than $D[i-1,j-1]$ by more than $GP$ (which has been defined as the cost of gap inserted aligned with a non-binding amino acid).

As it was already mentioned, $D[i,j]$ contains the minimum cost of transforming the $i$-length prefix of $seq_2$ to the $j$-length prefix of $seq_1$. If we know that the $i$-th character of $seq_2$ corresponds to a non-binding amino acid (indicated by '-'), it is always true that

$$D[i-1,j] - GP \leq D[i,j];$$

in other words, taking into account one more non-binding amino acid in an arbitrary sequence (in our case, $seq_2$) during DPA calculation can only reduce the previously calculated minimum by $GP$. This "reduction" only applies if the other sequence (in our case, $seq_1$) also contains a non-binding amino acid at position $j$.

For symmetry reasons, the equation above holds also if we swap the indexes $i$ and $j$ and write

$$D[i,j-1] - GP \leq D[i,j].$$

As these equations apply for any valid $i$ and $j$ indexes in the DPA matrix $D$ (except for the first and last rows and columns), we can express $D[i,j-1]$ and $D[i-1,j]$ in terms of $D[i-1,j-1]$:

$$D[i-1,j] - GP \leq D[i,j] \Longrightarrow D[i-1,j-1] - GP \leq D[i,j-1],$$

and similarly,

$$D[i,j-1] - GP \leq D[i,j] \Longrightarrow D[i-1,j-1] - GP \leq D[i-1,j].$$

Now we can state that our algorithm chooses the minimum of the following values:

$$\begin{cases} D[i-1,j-1] \\ D[i,j-1]+GP \geq (D[i-1,j-1]-GP)+GP = D[i-1,j-1] \\ D[i-1,j]+GP \geq (D[i-1,j-1]-GP)+GP = D[i-1,j-1] \end{cases}$$

This means that we do not have to calculate each cell located at the intersection of subsections consisting of non-binding amino acid sequences; instead, we can copy values diagonally.

Note that the first and last rows and columns of the DPA matrix $D$ cannot correspond to a non-binding amino acid in either sequences. This is due to the way of construction of the sequences that have to begin and end with a binding amino acid as described in Section 2.2.

If the amino acid sequences under comparison contain non-binding amino acids in a high proportion (which is usually the case), calculation of the defined distance can be speeded up significantly: intersections of two subsequences consisting of non-binding amino acids can be entirely skipped, copying values diagonally throughout the whole intersection.

The main principle behind the optimized algorithm presented here is somewhat similar to Ukkonen's approach [10]. Changes in the initialization and in the calculation of first and last rows and columns of the DPA matrix makes the presented algorithm unique. This is a consequence of taking amino acid sequences extracted from protein binding sites as an input and not simply arbitrary strings.

## 4  Runtime Measurements

Reduction in runtime is roughly proportional to the number of matrix cells located at the intersection of non-binding subsequences, compared to the size of the whole $D$ DPA matrix. However, to get a real picture about runtime improvement, it is necessary to apply the original and the optimized algorithm on a sample database consisting of real-world data. For this purpose we used sample databases consisting of protein chains obtained from 1000, 2000 and 5000 binding sites; all possible protein chain-pairs were compared to each other, resulting in approximately $\binom{1000}{2}$, $\binom{2000}{2}$ and $\binom{5000}{2}$ comparisons. The sample databases were constructed based upon the RS-PDB database. A few measurements regarding runtime improvement are given in Table 1.

**Table 1.** Runtimes of original and optimized versions of the algorithm(measured on a hardware containing a 3 GHz Pentium 4 processor)

| Sample DB size | Original algorithm | Optimized algorithm | Improvement factor |
|---|---|---|---|
| 1000 entries | 180 seconds | 36 seconds | 5.00 |
| 2000 entries | 654 seconds | 140 seconds | 4.67 |
| 5000 entries | 4340 seconds | 891 seconds | 4.87 |

## 5  Conclusion

Our main result is using binding residues for marking possibly significant amino acids in the sequence-given protein chain. All *binding amino acids* of a protein sequence are marked as *significant*; other amino acids are aggregated into one symbol. This special way of construction of amino acid sequences made the implementation of an optimized sequence comparison algorithm possible. Note, that we intend to apply this description only for analyzing the ligand-binding properties of protein structures.

A possible application might be the clustering of similar (in other words, low-distance) binding sites, or finding evolutionarily related sequences.

**Acknowledgments.** The author would like to thank Vince Grolmusz and Zoltán Szabadka their invaluable help. This research was partially supported by the European Commission FP6 program "scrIN-SILICO" and by the Hungarian OTKA agency, under grant No. T046234. Parts of this work were done in cooperation with Uratim Ltd. and Math-for-Health LLC.

## References

[1] Grolmusz, V., Szabadka, Z.: High Throughput Processing of the Structural Information of the Protein Data Bank. Journal of Molecular Graphics and Modeling 25, 831–836 (2007)

[2] Grolmusz, V., Szabadka, Z.: Building a Structured PDB: The RS-PDB Database. In: Proc. of the 28th IEEE EMBS Annual International Conference, New York City, pp. 5755–5758. IEEE Computer Society Press, Los Alamitos (2006)

[3] Henikoff, S., Henikoff, J.G.: Amino Acid Substitution Matrices from Protein Blocks. PNAS 89, 10915–10919 (1992)

[4] Mott, R.: Local Sequence Alignments with Monotonic Gap Penalties. Bioinformatics 15(6), 455–462 (1999)

[5] Needleman, S., Wunsch, C.: A general method applicable to the search for similarities in the amino acid sequence of two proteins. Journal of Molecular Biology 48(3), 443–453 (1970)

[6] Smith, T.F., Waterman, M.S.: Identification of Common Molecular Subsequences. Journal of Molecular Biology 147, 195–197 (1981)

[7] Altschul, S.F., Madden, T.L., Schaffer, A.A., Zhang, J., Zhang, Z., Miller, W., Lipman, D.J.: Gapped BLAST and PSI-BLAST: a new generation of protein database search programs. Nucleic Acids Research 27(17), 3389–3402 (1997)

[8] Berman, H.M., Henrick, K., Nakamura, H.: Announcing the worldwide Protein Data Bank. Nature Structural Biology 10(12), 980, PMID 14634627 (2003)

[9] Altschul, S.F., Gish, W., Miller, W., Myers, E.W., Lipman, D.J.: Basic Local Alignment Search Tool. Journal of Molecular Biology 215, 403–410 (1990)

[10] Ukkonen, E.: Algorithms for Approximate String Matching. Information and Control 64(1-3), 100–118 (1985)

# Comparing RNA Structures: Towards an Intermediate Model Between the EDIT and the LAPCS Problems

Guillaume Blin[1], Guillaume Fertin[2], Gaël Herry[2], and Stéphane Vialette[3]

[1] IGM-LabInfo - UMR CNRS 8049 - Université de Marne-la-Vallée
77 454 Marne-la-Vallée Cedex 2 - France
gblin@univ-mlv.fr
[2] LINA - FRE CNRS 2729 - Université de Nantes
44322 Nantes Cedex 3 - France
{fertin,herry}@lina.univ-nantes.fr
[3] LRI - UMR CNRS 8623 - Université Paris-Sud
91405 Orsay Cedex - France
vialette@lri.fr

**Abstract.** In the recent past, RNA structure comparison has appeared as an important field of bioinformatics. In this paper, we introduce a new and general intermediate model for comparing RNA structures: the Maximum Arc-Preserving Common Subsequence problem (or MAPCS). This new model lies between two well-known problems – namely the Longest Arc-Preserving Common Subsequence (LAPCS) and the EDIT distance. After showing the relationship between MAPCS, LAPCS, EDIT, and also the Maximum Linear Graph problem, we will investigate the computational complexity landscape of MAPCS, depending on the RNA structure complexity.

**Keywords:** RNA structures, arc-annotated sequences, motif extraction.

## 1 Introduction

In computational biology, the understanding of biological mechanisms is frequently induced by sequences comparison. However, in the context of RiboNucleic Acid molecules (RNA), one cannot focus only on sequences. Indeed, it is now clearly established that the conformation of an RNA molecule partially determines its function and therefore, RNA comparison has certainly to take into account both the sequence and the structure. From a combinatorial point of view, an RNA molecule may be described by the sequence of its bases *i.e.*, a single strand composed of the nucleotides A, C, G and U (also called the *primary structure*), together with the set of hydrogen bonds that connect pairs of bases. Those pairings induce a specific conformation, usually called *secondary structure* if it can be drawn planarly, and *tertiary structure* otherwise.

At a theoretical level, the RNA structure comparison problem has been addressed with different paradigms allowing flexibility on the comparison criteria

M.-F. Sagot and M.E.M.T. Walter (Eds.): BSB 2007, LNBI 4643, pp. 101–112, 2007.

[1,5,6,7,10,13,15]. Nevertheless, they all rely on the concept of *structure comparison*. In order to compare two RNA structures, one has to consider a set $\Delta$ of operations on bases and/or hydrogen bonds. Given such a set $\Delta$, comparing two RNA structures usually reduces to finding a series of operations of $\Delta$ – an *edit script* – that transforms one structure into the other. Providing a cost for each operation of $\Delta$ allows us to evaluate the *cost* $C$ of any edit script by summing the cost of all operations of the edit script. Then, referring to the standard *parsimony criterion*, the goal is to find the edit script transforming one structure into the other that minimizes the total cost $C$.

Existing paradigms mainly differ in the set of allowed operations: some consider only operations that can behave separately on bases and on hydrogen bonds, whereas others take into account operations that can act separately or simultaneously on bases and hydrogen bonds. In the past few years, essentially due to the increase of the number of determined RNA structures, their comparison has become all the more important. Unfortunately, for most paradigms, comparing such structures turns out to be an intractable problem. Nevertheless, recent research [11,16,18] has shown that relaxing the constraints on the preservation of the primary structure makes the RNA structure comparison problem tractable for more general cases including some types of tertiary structures. However, a certain gap lies between simplistic and sophisticated paradigms. The former ones only use the structure in order to constrain the possible edit scripts over a set of simple operations ; for instance, the conservation/loss of a hydrogen bond is not considered in the similarity computation. On the contrary, the latter ones consider more biologically relevant operations and their associated costs ; for instance, a simultaneous deletion of a hydrogen bond together with one or two of its incident bases is considered as a single operation, to which a specific cost is assigned. However, the existence of such operations in the model makes the problem become hard even for very restricted structures.

In this article, we introduce an intermediate paradigm, called Maximum Arc-Preserving Common Subsequence, or MAPCS. MAPCS uses the structure both in order to constrain the possible edit scripts and to estimate the similarity between two RNA structures, but with operations simpler than the ones of sophisticated paradigms (more precisely, MAPCS can be seen as a more realistic extension of the well-known LAPCS problem [10], while being simpler than EDIT). The reader should notice that MAPCS differs from the sequence-structure alignment problem defined by Bafna *et al.* [2] since arcs, that have to be preserved, add constraints on the possible edit scripts. After some preliminaries and definitions, we first describe how the MAPCS problem is related to the LAPCS and the MLG problem, introduced in [9]. We then study the computational complexity of the MAPCS problem. This study is another step towards establishing more precisely the complexity landscape of the RNA structure comparison problem.

## 2    Preliminaries and Related Works

From a combinatorial point of view, one can distinguish two types of modeling allowing for various flexibility and precision in the encoding of RNA structures:

(i) a representation that includes both the nucleotide sequence and the hydrogen bonds, called *arc-annotated sequence*, originally introduced by Evans [10], and (ii) representations using graphs, which do not necessarily take into account the precise label of the nucleotides that compose the sequence, such as 2-interval graphs [19] and linear graphs [9]. We will be more interested here in arc-annotated sequences and linear graphs (intensively studied in the recent past [1,5,6,7,10,13,15]).

## 2.1   Arc-Annotated Sequences : Problems EDIT and LAPCS

Given a finite alphabet $\Sigma$, an arc-annotated sequence is defined by a pair $(S_1, P_1)$, where $S_1$ is a string on $\Sigma^*$ and $P_1$ is a set of arcs connecting pairs of characters of $S_1$. In reference to RNA structures, we will refer to the characters as *bases*. The pair $(S_1, P_1)$ is called *RNA arc-annotated sequence* if $S_1 \in \{A, C, G, U\}^*$, and each arc of $P_1$ connects either bases A and U, or bases C and G. Any base with no arc incident to it is said to be *free*. Usually, five complexity levels reflecting the structure of the arcs are considered [10]: (1) PLAIN – there is no arc, (2) CHAIN – no base is incident to more than one arc and no two arcs are crossing or nest, (3) NESTED (NEST) – no base is incident to more than one arc and no two arcs are crossing, (4) CROSSING (CROS) – no base is incident to more than one arc and (5) UNLIMITED (UNLIM) – no restriction.

Those five levels respect an obvious inclusion relation denoted by the $\subset$ operator: PLAIN $\subset$ CHAIN $\subset$ NESTED $\subset$ CROSSING $\subset$ UNLIMITED. Notice that the absence of arcs makes PLAIN a very low informative level and, since PLAIN $\subset$ CHAIN, it is of little interest in the context of RNA structure comparison and will not be considered in this paper. In order to compare arc-annotated sequences, we consider the set of operations (and their associated costs) initially introduced in [17]. This set is composed of four substitution operations which induce renaming of bases in the arc-annotated sequence: *base-match* $(w_m : \Sigma^2 \rightarrow \mathbb{R})$, *base-mismatch* $(w_m : \Sigma^2 \rightarrow \mathbb{R})$, *arc-match* $(w_{am} : \Sigma^4 \rightarrow \mathbb{R})$, *arc-mismatch* $(w_{am} : \Sigma^4 \rightarrow \mathbb{R})$. Moreover, it also contains four deletion operations which induce deletion of bases and/or arcs, which we list together with their associated cost:

*base-deletion* $(w_d : \Sigma \rightarrow \mathbb{R})$   o-o-●-o-o → o-o—o-o

*arc-breaking* $(w_b : \Sigma^4 \rightarrow \mathbb{R})$   o-●-o⌒o-●-o → o-●-o⋯o-●-o

*arc-removing* $(w_r : \Sigma^2 \rightarrow \mathbb{R})$   o-●-o⌒o-●-o → o—o⋯o—o

*arc-altering* $(w_a : \Sigma^3 \rightarrow \mathbb{R})$   o-●-o⌒o-●-o → o—o⋯o-●-o or o-●-o⋯o—o

The *edit distance* between two arc-annotated sequences $(S_1, P_1)$ and $(S_2, P_2)$ is defined as the minimum cost of any edit script from $(S_1, P_1)$ to $(S_2, P_2)$. The problem consisting in finding this distance is called EDIT. To any edit script from $(S_1, P_1)$ to $(S_2, P_2)$ corresponds an *alignment* of the bases of $S_1$ and $S_2$ such that (i) any base which is inserted or deleted in a sequence is aligned with a *gap* (indicated by $-$) and (ii) any two bases (one per sequence) which are (mis)matched are aligned together. As illustrated in Table 1, Lin et al. proved in [17] that finding the edit distance between an arc-annotated sequence of CROSSING type and one of PLAIN type (denoted as EDIT(CROS, PLAIN)) is *MAX-SNP hard*. Thus, any harder problem (in terms of restriction levels) is also *MAX-SNP hard*. Moreover, they gave a polynomial-time dynamic programming algorithm for the problem EDIT(NEST, PLAIN). Blin et al. [6] showed that EDIT(NEST, NEST) is **NP**-complete.

The LAPCS problem was introduced by Evans in [10], and can be defined as follows: given two arc-annotated sequences $(S_1, P_1)$ and $(S_2, P_2)$, find the alignment of $S_1$ and $S_2$ which maximizes the number of matched positions and that satisfies the following conditions: for any arc $(i, j)$ in $P_1$ (resp. in $P_2$), if bases $i$ and $j$ are both matched to bases of $S_2$ (resp. $S_1$) – say $p$ and $q$ – then $(p, q)$ is an arc in $P_2$ (resp. $P_1$). In other words, an arc cannot be away from the alignment if none of its incident bases is away too. The computational complexity of the LAPCS problem has been studied in [10,15], and the main results are summarized in Table 1. Of importance here is the result of Blin et al. [7], who proved that the LAPCS problem can actually be seen as a very specific case of the EDIT problem. More precisely, LAPCS can be seen as a particular case of EDIT where the cost system for edit operations is the following: $w_r = 2w_d = 2w_a$, and all substitution operations and arc-breakings are prohibited with an arbitrary high cost. The main idea is to penalize deletion operations proportionally to the number of bases that are deleted.

## 2.2   Linear Graphs and the MLG Problem

As mentioned previously, one possible way of representing RNA structures is by means of linear graphs [9]. A linear graph of order $n$ is a vertex-labeled graph where each vertex is labeled by a distinct integer from $\{1, 2, \ldots n\}$ (the order of the vertices is induced by the labels) and is of degree at least one. Any edge between two vertices $i$ and $j$, with $i < j$, may be defined as the pair $(i, j)$. Linear graphs thus represent RNA structures in which only the hydrogen bonds are considered – the identity of the bases are ignored. Two edges of a linear graph are called *independent* if they do not share a vertex. Similarly to arc-annotated sequences, the four levels of arc structures CHAIN, NESTED, CROSSING and UNLIMITED may be used in this model.

In order to compare linear graphs, we define the notion of *occurrence* of one linear graph in another as follows. Given two linear graphs $G_1$ and $G_2$, $G_1$ is said to occur in $G_2$ (or $G_1$ is called a *subgraph* of $G_2$) if one can obtain $G_1$ from $G_2$ (regardless of the vertex labels) by a sequence of edge and vertex deletions. More formally, the deletion of vertex $i$ consists in (1) the deletion of all the edges incident to vertex $i$, (2) the deletion of vertex $i$ and of any vertex of degree zero and (3) the relabeling of all remaining vertices preserving the original order. Provided with those notations, the RNA structure comparison problem using linear graphs – noted MLG – is defined as follows: given two linear graphs $G_1$ and $G_2$, find a maximum size – in terms of edges – common linear subgraph. Hence, this problem comes down to finding a common substructure that has the

**Table 1.** RNA structure comparison: computational complexity of the LAPCS, EDIT and MLG problems considering input structures resp. of $A$ and $B$ types. For both the EDIT and the LAPCS problems, $n$ and $m$ denote resp. the number of bases of the sequences of $A$ and $B$ types. For the MLG problem, $n$ and $m$ denote the number of vertices of each linear graph, $n \geq m$.

| $A \times B$ | CHAIN | NESTED | | CROSSING | | | UNLIMITED | | | |
|---|---|---|---|---|---|---|---|---|---|---|
| | CHAIN | CHAIN | NEST | CHAIN | NEST | CROS | CHAIN | NEST | CROS | UNLIM |
| EDIT | $O(nm)$ [10] | $O(nm^3)$ [15] | NPC [6] | MAX-SNP hard [17] | | | | | | |
| LAPCS | $O(nm)$ [10] | $O(nm^3)$ [15] | | NPC [15,10] | | | | | | |
| MLG | $O(nm)$ [14] | $O(n^2m)$ [18] | $O(n^2m^2)$ [18] | $O(n^4 \log^3 n)$ [16] | NPC [8,19] | $O(n^4 \log^3 n)$ [16] | NPC [8,19] | | | |

largest number of arcs between two given structures. We note that there exists some variants of the problem [12,19].

As illustrated in Table 1, seeking for a maximal common substructure is easier when the maximality criterion relies only on the number of common arcs (MLG), rather than on common bases (LAPCS, EDIT). More precisely, one may note that when at least one of the input structures is of CHAIN or NESTED type, MLG is always polynomial time solvable.

## 3   Maximum Arc-Preserving Common Subsequence

In order to fill the gap which lies between simplistic and sophisticated paradigms like respectively, LAPCS and EDIT, we introduce here a new paradigm,that we name MAXIMUM ARC-PRESERVING COMMON SUBSEQUENCE. The purpose of MAPCS is to overcome both the lack of expressiveness of LAPCS and the intrinsic complexity of EDIT due to its sophisticated operations. Moreover, as illustrated in Table 1, according to the results of MLG, restricting the complexity of the similarity criteria may be a way of going further ; indeed, we then may be able to solve, in polynomial time, more instances. The MAPCS is defined formally as follows: given two arc-annotated sequences $(S_1, P_1)$ and $(S_2, P_2)$ and two functions $f_b : \Sigma \rightarrow \mathbb{N}^*$ and $f_a : \Sigma^2 \rightarrow \mathbb{N}^*$, find a common arc-annotated subsequence $(T, Q)$ that maximizes the following score function: $\sum_{c \in T} f_b(c) + \sum_{(c_1, c_2) \in Q} f_a(c_1, c_2)$. In other words, the MAPCS problem aims at finding a common subsequence similarly to the LAPCS problem, except that the score takes into account both the number of bases and arcs of the common subsequence. We will first focus on two possible extensions of the MAPCS problem, where either $f_a$ or $f_b$ always returns 0 ; for both problems, we state their computational complexity, depending on the form of the input structures (CHAIN, NEST, CROS or UNLIM). Then, we fully investigate the computational complexity of the MAPCS problem itself.

## 3.1   Extending MAPCS to the Case Where $f_a$ or $f_b$ Always Returns Zero

Before going into details concerning the computational complexity of the MAPCS problem, we would like to point out two closely related problems, that can be seen as extensions of MAPCS, if we allow function $f_a$ or $f_b$ to be ignored by always returning zero. In particular, we will see that those two problems are in fact closely related to, respectively, the LAPCS and the MLG problems.

If $f_a : \Sigma^2 \to 0$ and $f_b : \Sigma \to \mathbb{N}^*$, then the corresponding problem is equivalent to the LAPCS problem, whose complexity has been extensively studied, and is summarized in Table 1. If $f_a : \Sigma^2 \to \mathbb{N}^*$ and $f_b : \Sigma \to 0$, then the corresponding problem, that we will call MAPCS*, is closely related to the MLG problem, where the vertices of the linear graphs are now labeled with a letter from the alphabet $\Sigma = \{A, U, G, C\}$ and where edges only exist between two vertices labeled A and U (resp. C and G). The computational complexity of the MAPCS* problem, depending of the types of the input sequences, is summarized in the following propositions (some proofs are omitted due to space constraints).

**Proposition 1.** *The MAPCS*(CHAIN,CHAIN) problem can be solved in $O(nm)$ time, where $n$ and $m$ are the number of bases of each sequence.*

**Proposition 2.** *The MAPCS*(NEST,NEST) (resp. MAPCS*(NEST,CHAIN)) problem can be solved in $O(n^2 m^2)$ (resp. $O(nm^2)$) time, where $n$ is the number of bases of the NEST sequence and $m$ is the number of bases of the other sequence.*

*Proof.* Note that we can pre-process both sequences so that they do not contain free bases. Clearly, this pre-process does not change the result (sine $f_b$ always returns 0) neither the arc structure (i.e., CHAIN or NESTED), and can be carried out in linear time. Let us first focus on the MAPCS*(NEST,NEST) problem. The proof is directly derived from the work of Lozano and Valiente [18], in which a dynamic programming algorithm was given in order to obtain a maximum common embedded subtree of two trees. Briefly stated, since the two input sequences of our problem are of NESTED type, there is a natural representation of such sequences as trees: each vertex of the tree represents an arc, and an edge joins a father $f$ to its son $s$ if the arc represented by $s$ is directly nested in the arc represented by $f$. The dynamic programming algorithm from [18] computes the maximum common subtree of two trees, and thus, if adapted to the MAPCS* problem in order for the score function to take function $f_a$ into account, would output a common subsequence having the maximum score. The only thing missing in the algorithm from [18] is that it could match any two vertices of the tree (i.e., in our case, any two arcs). This means that, for instance, an arc A–U could be matched to an arc G–C. However, this can be easily fixed in the algorithm by adding some conditions, in the dynamic programming recursive formula, that will ensure that only similar arcs can be matched. It is possible to show that, using this algorithm to solve MAPCS*(NEST,CHAIN), the size of the dynamic programming table becomes $O(nm^2)$ and thus the time complexity follows. □

**Proposition 3.** MAPCS *(UNLIM,NEST) can be solved in $O(n^4 \log^3 n)$, where $n$ is the maximal number of bases in an input arc-annotated sequence.

*Proof.* The proof relies on the same argument as in proof of Proposition 2, using a result from the MLG problem [16]. The problem is to extract from two linear graphs of UNLIMITED type a NESTED structure having the maximum number of arcs. In our context, since one of the two input sequences is of NESTED type, we can ensure that the result will be NESTED. Moreover, in both problems, one wishes to maximize the number of arcs of the common structure. Thus, the algorithm from [16] could be applied to the MAPCS*(UNLIM,NEST) problem, except that linear graphs are unlabeled graphs, which means that any arc can be matched to any other arc in MLG. However, as for Proposition 2 above, this problem can be easily fixed: indeed, the algorithm from [16] starts by constructing trapezoids that correspond to all possible arc matchings, and then finds the maximum set of trapezoids that are either pairwise included, or totally disjoint, in order to end up with a NESTED structure of maximum size. In our case, it suffices to change the first step of the algorithm by constructing the trapezoids that correspond to "valid" matchings, that is arcs whose bases have the same labels in the same order. Hence the result.                                          □

**Theorem 1.** *The* MAPCS *(CROS,CROS) problem is* **NP**-*complete.*

*Proof.* We consider here the natural decision version of the MAPCS* problem, in the specific case where $f_a$ always returns the same constant. Clearly, the problem is in **NP**. In order to prove that it is **NP**-complete, we propose a polynomial reduction from the MAX-CLIQUE problem, defined as follows: Given a graph $G$ and an integer $k$, is there a clique of cardinality greater than or equal to $k$ in $G$ ? The idea here is to construct, from any graph $G = (V, E)$, two arc-annotated sequences $(S_1, P_1)$ and $(S_2, P_2)$, in which, informally, $(S_1, P_1)$ will represent $G$ and $(S_2, P_2)$ will represent a clique of cardinality $k$ that we wish to find in $G$. Then, we will prove that finding a common subsequence of maximum score between $(S_1, P_1)$ and $(S_2, P_2)$ is equivalent to finding a clique of size $k$ in $G$. Now, we formally describe the construction of the two arc-annotated sequences $(S_1, P_1)$ and $(S_2, P_2)$. We first describe $S_1$: $S_1 = S_1^1 S_1^2 ... S_1^n$, with $S_1^i = A(CG)^n U \; \forall i \in \{1, 2, ..., n\}$. Now, $P_1$ is defined as follows: first, within each $S_1^i$, there is an arc between bases A and U. Then, for each edge $(v_i, v_j)$ in $G$, we connect the $j$-th base C (resp. G) of $S_1^i$ to the $i$-th base G (resp. C) of $S_1^j$. Let us now describe the construction of $(S_2, P_2)$. We start with $S_2$: $S_2 = X_1 A Y_1 U X_2 A Y_2 U ... X_k A Y_k U X_{k+1}$, where (i) $X_i = (AU)^{n-k}$, and (ii) $Y_i = T_1 T_2 ... T_k$, with $T_j = CG$ for all $1 \leq j \leq k$. Thus, $S_2$ can be defined as $S_2 = ((AU)^{n-k} A(CG)^k U)^k (AU)^{n-k}$. The arcs of $P_2$ are as follows: each base A is connected by an arc to the first base U on its right. Moreover, for any $1 \leq i < j \leq k$, there is an arc between base C (resp. base G) in $T_j$ of $Y_i$ and base G (resp. base C) in $T_i$ of $Y_j$. Clearly, this construction can be achieved in polynomial time, and yields to sequences $(S_1, P_1)$ and $(S_2, P_2)$ that are both of CROSSING type. An illustration of such a construction is given in Figure 1, where $n = 4$ and $k = 3$.

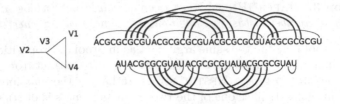

**Fig. 1.** Illustration of the construction where $n = 4$ and $k = 3$

The proof relies on the following equivalence: there exists a clique of cardinality greater than or equal to $k$ in $G$ iff there exists an arc-preserving common subsequence $(T, Q)$ of $(S_1, P_1)$ and $(S_2, P_2)$ whose score is greater than or equal to $(n + k(k - 1))f_a$. It is omitted here due to space constraints.                    □

**Table 2.** Complexity of MAPCS* ($n$ and $m$ are the lengths of the input sequences with $m \leq n$)

| $A \times B$ | CHAIN CHAIN | NESTED | | CROSSING | | | UNLIMITED | | | |
|---|---|---|---|---|---|---|---|---|---|---|
| | | CHAIN | NEST | CHAIN | NEST | CROS | CHAIN | NEST | CROS | UNLIM |
| MAPCS* | $O(nm)$ | $O(n^2m)$ | $O(n^2m^2)$ | $O(n^4 \log^3 n)$ | **NPC** | | $O(n^4 \log^3 n)$ | | **NPC** | |
| | Prop. 1 | Prop. 2 | Prop. 2 | Prop. 3 | Thm 1 | | Prop. 3 | | Thm 1 | |

Those results show that, though the problem is different, the MAPCS* and MLG problems are sufficiently close to admit the same computational complexity in each case (cf. Table 2).

### 3.2   The MAPCS Problem

We now turn to the MAPCS problem in itself, where neither $f_a$ nor $f_b$ returns zero. We first begin by a property relating MAPCS to LAPCS in a very specific case. This property will help us derive some computational complexity results.

*Property 1.* Given two arc-annotated sequences $(S_1, P_1)$ and $(S_2, P_2)$ s.t. at least one of them is of PLAIN type, then LAPCS and MAPCS have the same complexity.

*Proof.* Since at least one of $P_1$ and $P_2$ is the empty set, there will be no common arc in any common subsequence $(T, Q)$ of $(S_1, P_1)$ and $(S_2, P_2)$. Therefore, the score of $(T, Q)$ will only depend on $f_b$ for the MAPCS problem. Hence, the polynomial results for LAPCS can be adapted to MAPCS by changing the dynamic programming formulas to take $f_a$ into account, while the **NP**-completeness results for LAPCS actually are valid for MAPCS, when $f_b$ always returns 1.                    □

Thanks to Property 1, and since LAPCS(CROS,PLAIN) is **NP**-complete, so does MAPCS(CROS,PLAIN). This also implies that MAPCS(X,Y) is **NP**-complete, for any $X \in \{\text{CROS, UNLIM}\}$ and $Y$ s.t. $Y \subseteq X$.

**Proposition 4.** *The* MAPCS(CHAIN,CHAIN) *(resp.* MAPCS(NEST,CHAIN)) *problem is solvable in* $O(nm)$ *(resp.* $O(nm^3)$) *time, where* $n$ *(resp.* $m$*) is the number of bases of the* NEST *(resp.* CHAIN*) type sequence.*

*Proof.* Again, the idea here is to adapt an already existing algorithm, more precisely the dynamic programming algorithm designed by Lin et al. in [15] that was used to show that both the LAPCS(CHAIN,CHAIN) and the LAPCS(NEST,CHAIN) problems are polynomial-time solvable. Indeed, in [15], Lin et al. have designed a dynamic programming algorithm relying on a score function $\chi$ which may be refined to take into account the fact that the considered bases are incident or not to an arc ; in our case, it suffices to adapt $\chi$ in order to use either $f_b$ or $f_a$ in the computation of the score.                                                    □

**Theorem 2.** *The* MAPCS(NEST,NEST) *problem is* **NP***-complete.*

We consider here the natural decision version of MAPCS. We propose a reduction from the MIS-3P problem which is known to be **NP**-complete [4]. The MIS-3P problem consists in, given a cubic planar bridgeless connected graph $G = (V, E)$ and an integer $k$, finding an independent set of size $k$ in $G$. Recall, that a graph $G = (V, E)$ is said to be *cubic planar bridgeless connected* if any vertex of $V$ has degree three (cubic), $G$ can be drawn in the plane in such a way that no two edges of $E$ cross (planar), and there are at least two edge-disjoint paths connecting any pair of vertices of $G$ (bridgeless connected). As in [6], the proof is a two-step procedure: we first compute a 2-page book embedding of the input graph, and next transform each page into an RNA arc-annotated sequence.

A *2-page book embedding* of a graph $G$ is a linear ordering of the vertices of $G$ along a line together with an assignment of the edges of $G$ to the two half-planes delimited by the line – called the *pages* – such that no two edges assigned to the same page cross (they may, however, share a vertex). For convenience, we will refer to the page above (resp. below) the line as the *top-page* (resp. *bottom-page*). A *2-page s-embedding* will denote a 2-page book embedding where, in each page, every vertex has degree at least one.

**Theorem 3 ([3]).** *There exists a polynomial-time algorithm that computes a 2-page s-embedding of any cubic planar bridgeless connected graph.*

According to the above theorem, we thus first compute in polynomial-time a 2-page s-embedding of our input graph $G = (V, E)$, and we write $V = (v_1, v_2, \ldots, v_n)$ for the vertices of $G$ according to the linear ordering induced by the 2-page s-embedding. The corresponding NESTED type arc-annotated sequences $(S_1, P_1)$ and $(S_2, P_2)$ are defined as follows: $S_1 = X\ S_1^1\ X\ S_1^2 \ldots X\ S_1^n\ X$, $S_2 = X\ S_2^1\ X\ S_2^2 \ldots X\ S_2^n\ X$ where (i) $X = \mathsf{C}^{10n}\mathsf{G}^{10n}$ and there is an arc between the $i^{th}$ and the $20n - (i + 1)^{th}$ base of $X$, $1 \leq i \leq 10n$ (all bases of $X$ are thus paired in a nested way and we call all these arcs the *separating arcs* of the two arc-annotated sequences), and (ii) for each $1 \leq i \leq n$, $S_1^i$ (resp. $S_2^i$) is a segment AAAUUUAUAUA if vertex $v_i$ has degree 2 in the top-page (resp. bottom-page), and $S_1^i$ (resp. $S_2^i$) is a segment UAUAUAAAAUU otherwise ; moreover, in any segment AAAUUUAUAUA there is an arc between the $1^{st}$ (resp. $8^{th}$)

and the $5^{th}$ (resp. $9^{th}$) base, and in any segment UAUAUAAAAUU there is an arc between the $3^{rd}$ (resp. $7^{th}$) and the $4^{th}$ (resp. $11^{th}$) base ; we call all these arcs the *intra-segment arcs* of the two arc-annotated sequences.

What is left is to add the edges of the input graph $G$ into our construction. For each $(v_i, v_j) \in E$, $i < j$, of the top-page we create, in $P_1$, an arc $a_1$ linking a base U of $S_1^i$ and a base A of $S_1^j$ and an arc $a_2$, nested in $a_1$, linking the base A (resp. U) directly to the right (resp. left) of the base U (resp. A) of $a_1$. We proceed in a similar way for the bottom-page by adding, for each edge in that page, two arcs in $P_2$. Moreover, we impose that when a vertex $v_i$ has degree 1 in the top-page (resp. bottom-page), the two corresponding arcs in $P_1$ (resp. $P_2$) are incident to the two leftmost free bases A and U of the segment $S_1^i$ (resp. $S_2^i$), and to the four rightmost free bases A and U otherwise. We call all these arcs the *inter-segment arcs* of the arc-annotated sequences. It is easy to check that the above construction results in two NESTED type RNA arc-annotated sequences $(S_1, P_1)$ and $(S_2, P_2)$. An example of such a construction is given in Figure 2. The size of the sequences is clearly polynomial in $n$. Indeed, both $S_1$ and $S_2$ have length $20n^2 + 31n$. To complete the construction, we suppose that $f_a$ (resp. $f_b$) always returns the same constant. We claim that there exists an independent set of size $k$ in $G$ iff there exists an alignment of $(S_1, P_1)$ and $(S_2, P_2)$ with total score $20n(n + 1)f_b + 10n(n + 1)f_a + (f_a + 6f_b)k + \max\{6f_b, 5f_b + f_a\}(n - k)$.

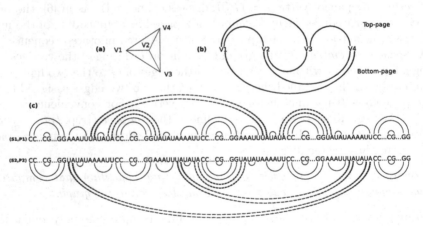

**Fig. 2.** (a) a cubic planar bridgeless connected graph $G$ of order 4, (b) a 2-page $s$-embedding of $G$ and, (c) the corresponding NESTED type arc-annotated sequences

The proof lies on several properties of an optimal alignment of the sequences $(S_1, P_1)$ and $(S_2, P_2)$ (Lemmas 1 to 4, whose proofs are omitted here). We call an alignment of $(S_1, P_1)$ and $(S_2, P_2)$ *canonical* if, for each $1 \leq i \leq n + 1$, the $i^{th}$ segment $X$ of $(S_1, P_1)$ is perfectly aligned to the $i^{th}$ segment $X$ of $(S_2, P_2)$.

**Lemma 1.** *Any optimal alignment of $(S_1, P_1)$ and $(S_2, P_2)$ is canonical.*

**Lemma 2.** *In any optimal alignment of $(S_1, P_1)$ and $(S_2, P_2)$, no inter-segment arc is conserved.*

We now consider the local alignment of two corresponding segments $S_1^i$ and $S_2^i$, $1 \leq i \leq n$. Clearly, $S_1^i = $ AAAUUUAUAUA and $S_2^i = $ UAUAUAAAAUU, or $S_1^i = $ UAUAUAAAAUU and $S_2^i = $ AAAUUUAUAUA. A simple calculation shows that the optimal alignment of $S_1^i$ and $S_2^i$ is of length 6 and preserves only one arc. Such an optimal alignment is obtained by deleting AAAUU in both $S_1^i$ and $S_2^i$ and has score $6f_b + f_a$. We refer to such an optimal alignment as an *optimal local alignment*. Furthermore, any non-optimal alignment of $S_1^i$ and $S_2^i$ results in a score at most $\max\{6f_b, 5f_b + f_a\}$.

**Lemma 3.** *The total score of any optimal alignment of* $(S_1, P_1)$ *and* $(S_2, P_2)$ *is* $20n(n+1)f_b + 10n(n+1)f_a + (f_a + 6f_b)k + \max\{6f_b, 5f_b + f_a\}(n - k)$, *where* $k$ *is the number of optimal local alignments.*

**Lemma 4.** *There exists an independent set of size* $k$ *in* $G$ *iff there exists an alignment of* $(S_1, P_1)$ *and* $(S_2, P_2)$ *with total score* $20n(n+1)f_b + 10n(n+1)f_a + (f_a + 6f_b)k + \max\{6f_b, 5f_b + f_a\}(n - k)$.

All the results concerning the MAPCS problem are summarized in Table 3.

**Table 3.** Complexity of MAPCS ($n$ and $m$ are the lengths of the input sequences with $m \leq n$)

| $A \times B$ | CHAIN | NESTED | | CROSSING | | | UNLIMITED | | | |
|---|---|---|---|---|---|---|---|---|---|---|
| | CHAIN | CHAIN | NEST | CHAIN | NEST | CROS | CHAIN | NEST | CROS | UNLIM |
| MAPCS | $O(nm)$ | $O(nm^3)$ | | NPC | | | | | | |
| | Prop. 4 | Prop. 4 | | Theorem 2 and Property 1 | | | | | | |

# 4  Conclusion

In this paper, we have introduced a new model for comparing two RNA structures using arc-annotated sequences – the MAPCS problem – which can be considered as an intermediate problem between LAPCS and EDIT. Indeed, it is less intricate than the EDIT problem, in the sense that some (but not all) of the edit operations have a specific cost. Moreover, it is a natural extension of the LAPCS problem in which a (non zero) score is given to any arc in the common subsequence, in addition to the score already given to its bases in the LAPCS problem. This new model makes RNA motif extraction biologically more relevant than the LAPCS problem, since one can intuitively think of a common RNA subsequence containing many arcs as more "reliable" than one containing as many bases, but less arcs. We have fully studied the computational complexity of the MAPCS problem. We have also found interesting to "locate" this problem compared to other well-known problems for RNA structure comparison, such as EDIT, LAPCS and MLG ; this allows us to show that MAPCS is not a mere extension of LAPCS, but somehow lies in the middle of those three problems. This new paradigm sheds light on a new aspect of the hardness of the RNA structures comparison problem ; namely, the hardness is not necessarily fully correlated to the complexity of the allowed edit operations.

# References

1. Alber, J., Gramm, J., Guo, J., Niedermeier, R.: Computing the similarity of two sequences with nested arc annotations. Th. Comp. Science 312(2), 337–358 (2004)
2. Bafna, V., Muthukrishnan, S., Ravi, R.: Computing similarity between RNA strings. In: Galil, Z., Ukkonen, E. (eds.) Combinatorial Pattern Matching. LNCS, vol. 937, pp. 1–16. Springer, Heidelberg (1995)
3. Bernhart, F., Kainen, P.C.: The book thickness of a graph. Journal of Combinatorial Theory, Series B 27(3), 320–331 (1979)
4. Biedl, T., Kant, G., Kaufmann, M.: On triangulating planar graphs under the four-connectivity constraint. Algorithmica 19, 427–446 (1997)
5. Blin, G., Fertin, G., Rizzi, R., Vialette, S.: What makes the arc-preserving subsequence problem hard ? In: Priami, C., Zelikovsky, A. (eds.) Transactions on Computational Systems Biology II. LNCS (LNBI), vol. 3680, pp. 1–36. Springer, Heidelberg (2005)
6. Blin, G., Fertin, G., Rusu, I., Sinoquet, C.: Extending the hardness of RNA secondary structure comparison. In: intErnational Symposium on Combinatorics, Algorithms, Probabilistic and Experimental methodologies (ESCAPE). LNCS, Springer, Heidelberg (to appear, 2007)
7. Blin, G., Touzet, H.: How to compare arc-annotated sequences: The alignment hierarchy. In: Crestani, F., Ferragina, P., Sanderson, M. (eds.) SPIRE 2006. LNCS, vol. 4209, pp. 291–303. Springer, Heidelberg (2006)
8. Bose, P., Buss, J.F., Lubiw, A.: Pattern matching for permutations. Information Processing Letters 65(5), 277–283 (1998)
9. Davydov, E., Batzoglou, S.: A computational model for RNA multiple structural alignment. Theoretical Computer Science 368(3), 205–216 (2006)
10. Evans, P.: Algorithms and Complexity for Annotated Sequences Analysis. PhD thesis, University of Victoria (1999)
11. Evans, P.A.: Finding common RNA pseudoknot structures in polynomial time. In: Lewenstein, M., Valiente, G. (eds.) CPM 2006. LNCS, vol. 4009, pp. 223–232. Springer, Heidelberg (2006)
12. Goldman, D., Istrail, S., Papadimitriou, C.H.: Algorithmic aspects of protein structure similarity. In: Found. of Comp. Science (FOCS), pp. 512–522 (1999)
13. Gramm, J., Guo, J., Niedermeier, R.: Pattern matching for arc-annotated sequences. In: Agrawal, M., Seth, A.K. (eds.) FST TCS 2002: Foundations of Software Technology and Theoretical Computer Science. LNCS, vol. 2556, pp. 182–193. Springer, Heidelberg (2002)
14. Hirschberg, D.S.: The longest common subsequence problem. PhD thesis, Princeton University (1975)
15. Jiang, T., Lin, G.-H., Ma, B., Zhang, K.: The longest common subsequence problem for arc-annotated sequences. In: Giancarlo, R., Sankoff, D. (eds.) CPM 2000. LNCS, vol. 1848, pp. 154–165. Springer, Heidelberg (2000)
16. Kubica, M., Rizzi, R., Vialette, S., Walen, T.: Approximation of RNA multiple structural alignment. In: Lewenstein, M., Valiente, G. (eds.) CPM 2006. LNCS, vol. 4009, pp. 211–222. Springer, Heidelberg (2006)
17. Lin, G.H., Ma, B., Zhang, K.: Edit distance between two RNA structures. In: Int. Conf. on Computational Biology (RECOMB), pp. 211–220. ACM Press, New York (2001)
18. Lozano, A., Valiente, G.: String Algorithmics. In: On the maximum common embedded subtree problem for ordered trees, ch. 7, pp. 155–169. King's College London Publications (2004)
19. Vialette, S.: On the computational complexity of 2-interval pattern matching problems. Theoretical Computer Science 312(2-3), 223–249 (2004)

# Evolving Phylogenetic Trees: A Multiobjective Approach

Guilherme P. Coelho[1], Ana Estela A. da Silva[2], and Fernando J. Von Zuben[1]

[1] Laboratory of Bioinformatics and Bioinspired Computing - LBiC
Department of Computer Engineering and Industrial Automation - DCA
School of Electrical and Computer Engineering - FEEC
University of Campinas - Unicamp
{gcoelho,vonzuben}@dca.fee.unicamp.br
[2] School of Mathematical and Nature Sciences - UNIMEP
aeasilva@unimep.br

**Abstract.** This work presents the application of the omni-aiNet algorithm - an immune-inspired algorithm originally developed to solve single and multiobjective optimization problems - to the construction of phylogenetic trees. The main goal of this work is to automatically evolve a population of phylogenetic unrooted trees, possibly with distinct topologies, by minimizing at the same time the minimal evolution and the mean-squared error criteria. The obtained set of phylogenetic trees contains non-dominated individuals that form the Pareto front and that represent the trade-off of the two conflicting objectives. The proposal of multiple non-dominated solutions in a single run gives to the user the possibility of having distinct explanations for the difference observed in the terminal nodes of the tree, and also indicates the restrictive feedback provided by the individual application of well-known algorithms for phylogenetic reconstruction that takes into account both optimization criteria, like Neighbor Joining.

## 1 Introduction

The phylogenetic tree reconstruction problem is interpreted here as a multiobjective problem, so that multiple objectives are considered simultaneously and the concept of a single optimal solution is no more applicable. In fact, to better explain the difference observed in the terminal nodes of the tree, the literature has provided a multitude of distinct optimization criteria that are generally applied in isolation [1]. Two of the most popular ones are the minimum evolution [2] and the mean-squared error criteria [3]. Both criteria can not be optimized simultaneously and the Neighbor Joining algorithm [4] is usually adopted to search for a phylogeny capable of optimizing them in an iterative way, thus producing a sub-optimal solution.

In this work, the multiobjetive optimization will be performed by an immune-inspired algorithm derived from the omni-aiNet algorithm [5], which is a high-performance population-based approach. Each individual in the population will

M.-F. Sagot and M.E.M.T. Walter (Eds.): BSB 2007, LNBI 4643, pp. 113–125, 2007.

be represented by an arbitrary distance matrix. This representation is adopted here to provide uniformity to the search space, and the genotype to phenotype mapping will be performed by the Neighbor Joining algorithm. In this way, instead of searching for proper topology and branch lengths, the search will be for a distance matrix that, when submitted to the Neighbor Joining algorithm, produces proper topology and branch lengths. In this genotype to phenotype mapping, the Neighbor Joining is only an instrument to construct a valid unrooted topology with branch lengths, and the distance matrices used as input will deserve no particular functional interpretation.

As already mentioned, the Neighbor Joining algorithm performs phylogenetic reconstruction in an iterative way, guiding to a single unrooted tree using a small amount of computational cost. That is why Neighbor Joining will be used both as a sub-optimal tool to find an unrooted tree with minimum evolution and minimum mean-squared error, and as a constructor of unrooted trees from arbitrary distance matrices. With this genotype representation, the Pareto front may contain non-dominated solutions with distinct topologies and also with the same topology but distinct branch lengths.

This work is divided as follows: in Section 2 some concepts of phylogenetic trees are presented. Basic definitions of multiobjective optimization and a brief explanation of the omni-aiNet algorithm are given in Section 3. Section 4 presents the adaptations made to the omni-aiNet to evolve phylogenetic trees. The experiments performed and the results obtained are given in Section 5, and finally Section 6 draws some concluding remarks.

## 2   Phylogenetic Trees

Philogeny or the history of the evolution of species is based on a concept of the Evolution Theory which asserts that groups of organisms that present similar attributes descend from a common ancestor. The main idea of the Evolution Theory is that all live beings have a certain degree of relation among each other.

Phylogenetic trees represent these evolutionary relations using relationships among species. They are constructed based on character data, being a character any characteristic of an organism that can assume different states. A typical biological example of a character is the position of a nucleotide in a DNA sequence, being the state of the character the nucleotide itself (A,C,G,T). Figure 1 illustrates two phylogenetic trees of DNA sequences.

A phylogenetic tree can be classified into several forms [1]. The two most common ways of classification are: rooted and unrooted trees, and multifurcating and bifurcating trees (which is related to the number of edges that originates from a given node). A rooted tree is a tree that has only one common ancestor for all species, which is called the root of the tree (Figure 1(b)). Consequently, an unrooted tree is a tree that does not establish a direction of evolution (Figure 1(a)). In phylogeny, bifurcating trees are generally adopted, indicating that each species evolves generating two descending species. In this work, only bifurcating unrooted trees will be considered.

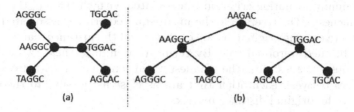

**Fig. 1.** Phylogenetic trees of DNA sequences: (a) an unrooted and (b) a rooted tree

It is important to highlight that, even though Figures 1 (a) and (b) present known sequences associated with internal nodes, they are generally considered as hypothetical nodes, because only terminal nodes have known sequences.

## 2.1  Classical Approaches to Tree Reconstruction

The reconstruction of phylogenetic trees can be made using several computational procedures, that can be classified as follows: (1) algorithmic: a systematic procedure is designed to determine the tree; (2) search-based: each tree corresponds to a point in a search space to be explored and an optimization criterion is defined to compare alternative proposals of tree topologies. The algorithmic procedures try to simultaneously perform the definition of the topology and the fulfillment of the optimization criteria, in a greedy and step-by-step manner, while the search-based procedures adopt a distinct paradigm, based on computational search strategies and composed of two steps: one for the evaluation of topologies, taking evolutionary assumptions, and other for the determination of topologies with the highest evaluation among the candidates.

The main disadvantages of the algorithmic procedures are the possibility of getting stuck in poor local minima and the presentation of a single tree topology at the end of its execution. The main advantage is the reduced computational cost, with the number of hypothetical ancestors to be determined being linearly associated with the number of entities under analysis. These methods incorporate all distance-based approaches, including cluster analysis (e.g. UPGMA) and Neighbor Joining (NJ).

The opposite scenario can be depicted in the case of search-based procedures. The computational cost is the main disadvantage, due to the factorial increase in the number of candidate topologies with the number of entities under analysis. On the other hand, very desirable aspects are the possibility of avoiding poor local minima along the search and the capability of presenting several high-quality tree topologies as the output, instead of a single one.

In this work, we explore the advantages of an algorithmic procedure for tree reconstruction (the Neighbor Joining, to be described in this section) together with an immune-inspired algorithm, originally developed to perform multiobjective optimization (the omni-aiNet algorithm, to be described in Section 3), to evolve a population of phylogenetic trees, trying to minimize at the same time the minimum evolution and the mean-squared error from a given distance matrix.

The minimum evolution criterion is associated with the sum of the lengths of all the branches of the tree, while the mean-squared error is associated with the difference between the original distance matrix and the distance matrix extracted from the obtained unrooted tree. By trying to minimize these two criteria, we intend to obtain trees with the smallest total length and with taxa presenting distances between each other that are as close as possible to the distances provided by the original distance matrix.

**The Neighbor Joining Algorithm:** The Neighbor Joining algorithm is a method for reconstructing unrooted phylogenetic trees taking a matrix of evolutionary dissimilarities as input data (referred in this text simply as *distance matrix*). The dimension of this square matrix of pairwise distances corresponds to the number of leaves in the resulting unrooted tree topology, each one denoted a taxon or OTU (operational taxonomic unit). Initially, the $n$ taxa are neighbors, because the algorithm starts with a star tree. A sequence of agglomerative steps then follows, taking into account the minimum evolution principle [2] to determine the pair of taxa to be joined, among all the $n * (n - 1)/2$ possibilities, and the Fitch-Margoliash approach [6] to propose the branch lengths of the two new branches, trying to minimize the mean squared error. At each agglomerative step, a new node is created (HTU - hypothetical taxonomic unit) to support the two additional branches, so that the star tree looses the newly-joined OTUs and gains the new HTU in replacement. This iterative process is repeated until the remaining star tree has only three taxa.

In computational terms, the agglomerative process has a computational complexity of $O(n^3)$, where n is the number of taxa, and may be interpreted as a greedy strategy that tries to simultaneously satisfy the minimum evolution principle [2], associated with the sum of branch lengths, and the least squares criterion [3], associated with the difference between the original distance matrix and the distance matrix extracted from the obtained unrooted tree. As a consequence, taking locally best decisions toward optimizing both objectives can not guarantee in general the achievement of the global minimum evolution tree, but only a sub-optimal tree whose topology may be similar (and sometimes identical) to the minimum evolution tree [7].

Due to the greedy nature of the search, among all the candidate pairs at each step of the agglomerative process, only one candidate is taken. As a consequence, the NJ algorithm is only capable of producing a single unrooted tree at the end of the execution.

## 2.2   Distance Metrics

Once the algorithms for phylogenetic reconstruction may produce several different trees as the output, a technique to compare those resulting trees is necessary. So, several techniques for measuring distance between trees were proposed in the literature ([1], [8], [9]), as the *Nearest Neighbor Interchange* [10], *Quartet Distance* [11] and the most popular *Robinson-Foulds* metric [12], which was adopted in this work.

The Robinson-Foulds metric is also known as *symmetric difference* or *partition metric*, and consists in dividing an unrooted phylogenetic tree into two partitions. In order to divide the tree, any branch or edge may be chosen, and this edge divides the original tree into two new trees, each one connected to a final point of the chosen edge. Each edge on a tree points out a specific partition of the original tree.

The symmetric difference between two distinct trees consists of the sum of the absolute differences between the corresponding edge lengths in the trees under comparison. When a given edge exists in one tree and does not exist in the other, the length of the absent edge is considered as "zero".

As will be seen in Section 4, the Robinson-Foulds metric was used in this work in the suppression phase of the omni-aiNet algorithm (see Section 3.2), where the phenotype of the individuals in the population are compared with each other.

# 3   Multiobjective Optimization

In the last two decades, evolutionary computation has been successfully applied to multiobjective optimization problems, what leaded to a new research field, namely EMO (Evolutionary Multiobjective Optimization [13], [14]). Many algorithms specialized to this kind of problems were proposed (e.g. [15], [16], [17]), each of them with its own characteristics and particular mechanisms.

In this section, we will formalize a multiobjective problem and give definitions of some concepts commonly adopted in multiobjective optimization, that are relevant to this work. Subsequently, the omni-aiNet algorithm [5] will be described.

## 3.1   Basic Concepts

A multiobjective optimization problem (MOP) is a problem which has two or more objectives that must be optimized simultaneously. This kind of problem generally presents constraints imposed on the objectives and on the domain of the variables, and also objectives that can be in conflict with each other. These conflicts among the objectives often lead to a set of non-dominated solutions for the problem (which represent good compromises among the objectives), instead of a single optimal solution.

A given vector $u = (u_1, \ldots, u_k)$ is said to dominate a vector $v = (v_1, \ldots, v_k)$ (denoted by $u \preceq v$) if and only if all $k$ components of $u$ are better or equal to the corresponding components of $v$ and there is at least one component of $u$ that is strictly better than the corresponding component of $v$.

When we have a single objective $f$, the optimal solution corresponds to the point that has the smallest value of $f$, considering the whole search space (in a minimization problem). However, for several objective functions, the notion of "optimal" solution changes, because the aim now is to find good trade-offs among the objective functions [1]. In this case, the most commonly adopted notion

---

[1] If the objective functions are not conflicting, a single solution exists for the MOP.

of optimality is the one associated with the *Pareto front*. A solution $x^*$ belongs to the Pareto front if there is no other feasible solution $x$ (i.e. a solution that does not violate any constraints) capable of reducing the value of an objective without simultaneously increasing at least one of the others.

Therefore, the *Pareto optimal set* (the optimal solution) for a multi-objective optimization problem is given by the set of solutions that is not dominated by any other feasible solution in the domain of the problem, and the corresponding *Pareto front* of this optimal set is the set obtained by the application of the objective functions to each solution in the *Pareto optimal set*. An example of *Pareto front* can be found in Figure 4.

## 3.2   The Omni-aiNet Algorithm

The omni-aiNet is an immune-inspired algorithm proposed by Coelho and Von Zuben [5] to solve single and multi-objective optimization problems, either with single and multi-global solutions. The search engine of this algorithm is capable of automatically adapting the exploration of the search space according to the intrinsic demand of the optimization problem. Due to the immune inspiration, the omni-aiNet presents a population capable of adjusting its size during the execution of the algorithm, according to a predefined suppression threshold, and a grid mechanism to control the spread of solutions in the objective space (space of the objective functions).

The omni-aiNet algorithm works with a real-coded population of antibodies that correspond to the candidate solutions for the optimization problem, and basically follows the steps shown in Figure 2.

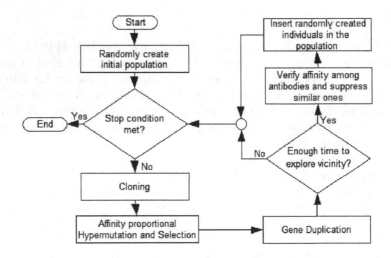

**Fig. 2.** Main steps of the omni-aiNet

The algorithm starts by randomly generating an initial population of size $N_i$ ($N_i$ is defined by the user). Each individual generated is within the range of the variables. After the creation of the initial population, the algorithm enters a loop where the stop criterion is the number of generations (also defined by the user). Within this loop, the main steps of the algorithm are executed: Cloning, Hypermutation (using a *polynomial mutation* mechanism), Selection and Gene Duplication (duplication of parts of the elements in the DNA chain of an individual). The suppression of individuals and insertion of new randomly generated ones are made from $N_{gs}$ to $N_{gs}$ generations ($N_{gs}$ is defined by the user). The value of $N_{gs}$ should be greater than one to give enough time for the algorithm to explore the vicinity of each solution before the suppression of similar individuals.

Further details about each mechanism of the omni-aiNet algorithm can be found in [5] and will be omitted here. In the next section, the modifications implemented into the original omni-aiNet algorithm, aiming at evolving phylogenetic trees, will be described in detail.

## 4    The Omni-aiNet Applied to Phylogenetics

As mentioned before, in this work we have applied the omni-aiNet algorithm to evolve a population of phylogenetic trees of possibly different topologies. To do so, three dedicated modules were incorporated: (*i*) an adequate encoding of the individuals by means of a genotype to phenotype mapping; (*ii*) a mechanism to generate the initial population; and (*iii*) a mechanism to compare similar individuals in the suppression phase of the algorithm (see Figure 2). With these dedicated modules, no further adaptations to the algorithm should be made to effectively evolve a set of phylogenetic trees.

In this section, we will outline these dedicated modules and describe the feasibility constraints applied to the individuals in the population.

### 4.1    Encoding the Trees

As can be seen in Section 2.1, the Neighbor Joining algorithm is an effective method to obtain a phylogenetic tree from a given distance matrix. In this work, we explore this characteristic by working directly with distance matrices as individuals in the population (the *genotype* of the individuals), and using the Neighbor Joining to convert these matrices into the corresponding trees (the *phenotype* of the individuals).

This approach allows us to maintain the real encoding used by the original algorithm and, at the same time, indirectly permits the evolution of trees with distinct topologies without the need of refined data structure devices to manipulate trees.

A brief illustration of the enconding mechanism and the conversion of the encoded individual (genotype) to the corresponding phylogenetic tree (phenotype) is given in Figure 3.

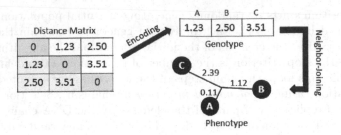

**Fig. 3.** Illustration of the encoding mechanism and the conversion from distance matrix to phylogenetic tree using Neighbor Joining

## 4.2 Initial Population

In the original algorithm, the initial population of $N_i$ individuals ($N_i$ defined by the user) is generated randomly within the domain of the problem. We could adopt the same approach here. However, since we have an initial distance matrix defined for the taxa of the problem, we have decided to initialize the individuals by adding small random perturbations to this original matrix. By doing so, the initial individuals tend to be closer to the original distance matrix than a purely random initial population would be. Notice again that the distance matrices in the population have no relevant meaning and should serve only as a proper input to the Neighbor Joining algorithm. We are looking for distance matrices that, when used as input to the Neighbor Joining algorithm, produces high quality unrooted trees at the output.

Therefore, the application of the omni-aiNet to generate a set of distance matrices that lead to high quality trees when submitted to the NJ algorithm can be seen as the result of the application of perturbations (proportional to each individual's quality, as can be seen in [5]) to the original distance matrix, that can compensate the NJ's possibly unsatisfactory results obtained when directly applied to this original distance matrix.

## 4.3 Affinity Among Antibodies and Suppression

The measure of affinity among antibodies (individuals in the population) together with the suppression phase of the omni-aiNet algorithm is one of the key mechanisms to maintain a good diversity of the individuals in the population (the second diversity maintenance mechanism in the algorithm is the *grid* procedure). The maintenance of diversity in the population of candidate solutions is important to allow the algorithm to perform a wide exploration of the search space and, consequently, increase the overall efficiency of the search, and provide a good spread of the final solutions along the Pareto front.

In the original algorithm, the measure of affinity among antibodies (that is nothing more than a measure of similarity of a pair of individuals) is given by the Euclidean distance from one antibody to the other. In this work, the Euclidean distance of distance matrices is not directly related to the dissimilarity among the corresponding phylogenetic trees, what could produce misleading results. To avoid this problem, we have replaced the Euclidean distance by the Robinson-Foulds metric (see Section 2.2), that is capable of evaluating how distinct a given tree is from another one. Once this metric is symmetrical and also returns real values, it can simply replace the Euclidean distance in omni-aiNet without the need of deeper modifications into the original algorithm.

### 4.4  Feasibility of Solutions

One last topic that should be highlighted is the feasibility of the candidate solutions. In the original omni-aiNet, developed for function optimization, to define the feasibility of a candidate solution it is verified whether it satisfies the equality, inequality and domain constraints of the problem. In this work, the procedure is the same, except that we don't have any equality constraints and we must analyze not only the genotype of the individuals (their distance matrices) but also its phenotype (the resulting phylogenetic trees).

If we apply only the restriction of positivity to the values in the distance matrices (once it does not make sense any negative distances) we do not guarantee that the branch lengths of the trees obtained by the Neighbor Joining will also be non-negative (see Equation 6 in [4]). Negative branches can appear due to noisy values in the distance matrix which can possibly lead to non-additive distances in the matrices [18]. So, we should also analyze the branch lengths of an individual to determine its feasibility.

In summary, we adopted two constraints to the problem: the values in the distance matrices should be positive (and we also adopted an upper bound, as required by the algorithm) and the values of the edge lengths of the phylogenetic tree associated with each individual should also be non-negative.

## 5  Experimental Results

Although several practical aspects of the multiobjective algorithm are of concern, in this paper we will emphasize the strong contrast between the results produced by the isolated application of Neighbor Joining and the ones produced by the omni-aiNet algorithm. Notice that both approaches take into account minimum evolution and mean squared error as optimization criteria. A single instance of a distance matrix associated with eight taxa is considered, given in Equation 1. Even in this small-size phylogenetic problem the performance of the Neighbor Joining will be shown to be poor, because the sub-optimal solution is located far from the Pareto front produced by the multiobjective approach.

In Subsection 5.1 the parameters of the omni-aiNet will be presented and the values attributed to them in this experiment will be given, and in Subsection 5.2 the obtained results will be presented and discussed.

$$D = \begin{pmatrix} 0 & 7 & 8 & 11 & 13 & 10 & 13 & 17 \\ 7 & 0 & 5 & 8 & 10 & 8 & 10 & 14 \\ 8 & 5 & 0 & 5 & 7 & 10 & 7 & 11 \\ 11 & 8 & 5 & 0 & 8 & 5 & 8 & 12 \\ 13 & 10 & 7 & 8 & 0 & 1 & 6 & 10 \\ 10 & 8 & 10 & 5 & 1 & 0 & 9 & 13 \\ 13 & 10 & 7 & 8 & 6 & 9 & 0 & 8 \\ 17 & 14 & 11 & 12 & 10 & 13 & 8 & 0 \end{pmatrix} \tag{1}$$

## 5.1   Parameters

The omni-aiNet has several parameters that should be adjusted before the execution of the algorithm. These parameters will be briefly described here, together with the respective values adopted in this work. All parameters were chosen in a way to lead the algorithm to present a good performance. However, no exhaustive exploration of the parameter space was made. Further information about these parameters can be obtained in [5].

- **Size of the initial population:** controls the number of individuals that will be generated initially. In this work we set this parameter to 20 individuals.
- **Maximum number of iterations:** this is the stop criterion of the algorithm. We fixed this parameter in 50 iterations.
- **Number of random individuals inserted:** indicates the number of randomly generated individuals that will be inserted in the population after the suppression phase. This parameter was set to 10 individuals.
- **Number of clones for each individual:** corresponds to the number of clones that are generated for each individual in the population, during the clonal expansion of the algorithm. We defined this parameter as 10 clones per individual.
- **Suppression Threshold:** determines whether an individual must be suppressed (eliminated) from the population or not. In the suppression phase of the algorithm, all individuals are compared to the remaining ones and, if their affinity is below a given threshold, the one with the smallest fitness is suppressed. In this work the suppression threshold was set to 0.05, what means that if two individuals have a distance value between each other smaller than 5% of the distance between the farthest individuals in the population, the worst one will be suppressed.
- **Maximum population size:** this parameter is a limit to the number of individuals in the population and was set to 100 individuals
- **$\delta$:** controls the mechanism of $\epsilon$-dominance (a relaxation of the Pareto dominance concept), and stimulates more exploration of the search space, improving diversity. This parameter was set to 0.005.

## 5.2   Results

The final set of non-dominated solutions, obtained by the omni-aiNet is presented in Figure 4 (circles), together with the phylogenetic tree obtained by the Neighbor Joining method applied to the original distance matrix of the problem

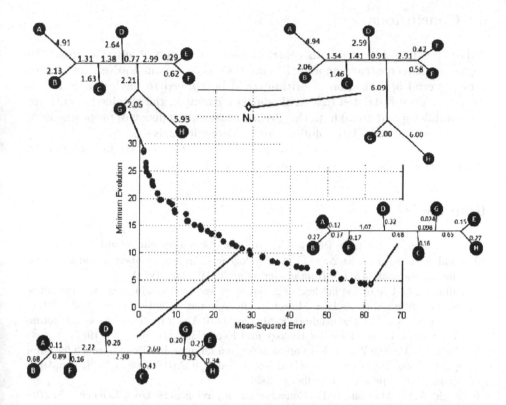

**Fig. 4.** Final non-dominated solutions found by the omni-aiNet algorithm (circles) and the tree obtained by the Neighbor Joining for the original distance matrix (diamond). This figure also presents three of the final trees obtained by the omni-aiNet algorithm. These results were obtained after 7 hours of simulation on a Pentium IV 2.6GHz, with 1.5GB of RAM, using the Matlab environment.

(diamond). As can be seen, the solution obtained by the Neighbor Joining is worse than the ones obtained by the omni-aiNet (as it is positioned farther from the Pareto front), which demonstrates the capability of the proposed methodology to generate phylogenetic trees with better mean-squared error and minimum evolution. These results also clearly illustrate the drawback of the greedy and iterative optimization of the two objectives made by the Neighbor Joining, once the solution found by this algorithm is sub-optimal.

Figure 4 also presents the graphical representation of three of the phylogenetic trees generated by the omni-aiNet and, as can be seen, the proposed methodology is indeed capable of evolving trees with distinct topologies, and also trees with the same topology but distinct branch lengths. It is interesting to note that the omni-aiNet was capable of obtaining a tree with the same topology obtained by the Neighbor Joining, but with different branch lengths.

## 6   Conclusions

This paper presents two main contributions: (*i*) a multiobjetive approach to phylogenetic reconstruction; and (*ii*) adaptations to the omni-aiNet so that this very powerful optimization algorithm could be applied to phylogenetic reconstruction. Even restricted to a single didactic example, the obtained results are very promising and are akin to the already observed tendency of proposing multiple alternative candidate solutions in phylogenetic analysis.

As further perspectives, we intend to consider datasets with distinct aspects and to introduce additional objectives to be optimized.

## References

1. Felsenstein, J.: Inferring Phylogenies. Sinauer Associates, Suderland, USA (2004)
2. Kidd, K.K., Sgaramella-Zonta, L.A.: Phylogenetic analysis: Concepts and methods. The American Journal of Human Genetics 23, 235–252 (1971)
3. Bulmer, M.: Use of the method of generalized least squares in reconstructing phylogenies from sequence data. Molecular Biology and Evolution 8, 868–883 (1991)
4. Saitou, N., Nei, M.: The neighbor-joining method: A new method for reconstructing phylogenetic trees. Molecular Biology and Evolution 4(4), 406–425 (1987)
5. Coelho, G.P., Von Zuben, F.J.: omni-aiNet: An immune-inspired approach for omni optimization. In: Bersini, H., Carneiro, J. (eds.) ICARIS 2006. LNCS, vol. 4163, pp. 294–308. Springer, Heidelberg (2006)
6. Fitch, W.M., Margoliash, E.: Construction of phylogenetic trees. Science 155, 279–284 (1967)
7. Saitou, N., Imanishi, M.: Relative efficiencies of the fitch-margoliash, maximum-parsimony, maximum-likelihood, minimum-evolution, and neighbor-joining methods of phylogenetic reconstructions in obtaining the correct tree. Molecular Biology and Evolution 6, 514–525 (1989)
8. Brodal, G.S., Fagerberger, R., Pedersen, C.N.S.: Computing the quartet distance between evolutionary trees in time $O(n.log(n))$. Algorithmica 38, 377–395 (2004)
9. DasGupta, B., He, X., Jiang, T., Li, M., Tromp, J., Zhang, L.: On distances between phylogenetic trees. In: Proceedings of the 8th Annual ACM - SIAM Symposium on Discrete Algorithms, pp. 427–436. ACM Press, New York (1997)
10. DasGupta, B., He, X., Jiang, T., Li, M., Tromp, J., Zhang, L.: On computing the nearest neighbor interchange distance. Mathematics Subject Classification (1991)
11. Bryant, D.: A classification of consensus methods for phylogenetics. In: Janowitz, M.F., Lapoint, F.J., Morris, F.R., Mirkin, B., Roberts, F.S. (eds.) Bioconsensus. Dimacs Series in Discrete Mathematics and Theoretical Computer Science, vol. 61, American Mathematical Society (2003)
12. Robinson, D.F., Foulds, L.R.: Comparison of phylogenetic trees. Mathematical Biosciences 53, 131–147 (1981)
13. Coello Coello, C.A., Van Veldhuizen, D.A., Lamont, G.B.: Evolutionary Algorithms for Solving Multi-Objective Problems. Kluwer Academic Publishers, New York (2002)
14. Deb, K.: Multi-Objective Optimization using Evolutionary Algorithms. John-Wiley & Sons, Chichester, UK (2001)

15. Corne, D.W., Jerram, N.R., Knowles, J.D., Oates, M.J.: PESA-II: region-based selection in evolutionary multiobjective optimization. In: Proceedings of the Genetic and Evolutionary Computation Conference (GECCO-2001) (2001)
16. Deb, K., Pratap, A., Agarwal, S., Meyarivan, T.: A fast and elitist multiobjective genetic algorithm: NSGA-II. IEEE Transactions on Evolutionary Computation 6(2), 182–197 (2002)
17. Zitzler, E., Laumanns, M., Thiele, L.: SPEA2: Improving the strength pareto evolutionary algorithm. In: EUROGEN 2001. Evolutionary Methods for Design, Optimization and Control with Applications to Industrial Problems, pp. 95–100 (2002)
18. Atteson, K.: The performance of neighbor-joining methods of phylogenetic reconstruction. Algorithmica 25, 251–278 (1999)

# Comparing Several Approaches for Hierarchical Classification of Proteins with Decision Trees

Eduardo P. Costa[1], Ana C. Lorena[2], André C. P. L. F. Carvalho[1],
Alex A. Freitas[3], and Nicholas Holden[3]

[1] Depto. Ciências de Computação
ICMC/USP - Sao Carlos - Caixa Postal 668
13560-970 - Sao Carlos-SP, Brazil
{ecosta,andre}@icmc.usp.br
[2] Universidade Federal do ABC
09.210-170 - Santo André-SP, Brazil
ana.lorena@ufabc.edu.br
[3] Computing Laboratory
University of Kent, Canterbury, CT2 7NF, UK
{a.a.freitas,nh56}@kent.ac.uk

**Abstract.** Proteins are the main building blocks of the cell, and perform almost all the functions related to cell activity. Despite the recent advances in Molecular Biology, the function of a large amount of proteins is still unknown. The use of algorithms able to induce classification models is a promising approach for the functional prediction of proteins, whose classes are usually organized hierarchically. Among the machine learning techniques that have been used in hierarchical classification problems, one may highlight the Decision Trees. This paper describes the main characteristics of hierarchical classification models for Bioinformatics problems and applies three hierarchical methods based on the use of Decision Trees to protein functional classification datasets.

## 1 Introduction

In functional genomics, an important problem is the prediction of the function of proteins. Proteins are the main building blocks of the cell, and perform almost all the functions related to cell activity. The primary sequence of a protein consists of a linear string of amino acids, which is then folded into a specific 3-D shape necessary for the protein to function properly. Proteins often share common amino acid sub-sequences due to evolutionary processes.

An approach frequently used in the prediction of a protein function is to search for similar sequences in protein databases. The objective is to find a similar sequence whose function is known. If a similar protein sequence is found, its function is assigned to the new protein. Although this method is very useful in a large number of situations, it has also some limitations [1]. Two proteins might have very similar sequences and perform different functions, or have very different sequences and perform the same or a similar function. Additionally, the

M.-F. Sagot and M.E.M.T. Walter (Eds.): BSB 2007, LNBI 4643, pp. 126–137, 2007.

proteins being compared may be similar in regions of the sequence that are not determinants of their function.

A second approach may be used alternatively or in complement to the similarity-based approach. The central idea of this approach consists of inducing a classification model for the prediction of protein function. Each protein is represented by an attribute set and a learning algorithm captures the most important relationships between the attributes and the classes present in the dataset.

As protein functional data is, frequently, organized hierarchically (for example, in the Gene Ontology [2] and in the Enzyme Commission hierarchy [3]), the use of hierarchical techniques for the induction of classification models in Bioinformatics is a promising research area.

This paper treats the main aspects concerned with hierarchical classification in Bioinformatics. Two datasets were used for a comparative study among different schemes for hierarchical classification. Decision Trees (DTs) [4] were used in the classifiers induction.

The main contribution of this paper is to compare several different approaches for the hierarchical classification of proteins. To the best of our knowledge, an empirical comparison of the approaches evaluated in this paper has not been reported yet in the literature.

The paper is organized as follows: Section 2 introduces important concepts of hierarchical classification; Section 3 presents the main approaches used in the induction of classifiers for hierarchical classification problems, as well as some literature related to the protein function prediction problem; Section 4 discusses the materials and methods employed in the experiments performed in this work; Section 5 presents the experimental results; and Section 6 has the main conclusions of this work.

## 2  Hierarchical Classification

Classification is one of the most important problems in Machine Learning (ML) and Data Mining (DM) [5]. A classification problem can be defined as the the the process of finding a function, through a training or adjustment phase, which maps each input instance $T_i$ into one of the $N$ classes of the problem, with $i = 1, 2, ..., n$, where $n$ is the number of training instances.

The vast majority of classification problems reported in the literature involves flat classification, where each instance is assigned to a class out of a finite (and usually small) set of flat classes. Nevertheless, there are more complex classification problems, where the classes to be predicted are hierarchically related [1,6]. In these classification problems, one or more classes can be divided into subclasses or grouped into superclasses. These problems are known as hierarchical classification problems in the ML literature.

There are two main ways in which the classes may be hierarchically disposed: as a tree or as a Directed Acyclic Graph (DAG). The main difference between the tree structure (Figure 1.a) and the DAG structure (Figure 1.b) is that, in the tree structure, each node has just one parent node, while in the DAG each node may have more than one parent. For both flat and hierarchical classification schemes, the nodes represent the problem classes and the root node corresponds to "any class", denoting a total absence of knowledge about the class of an object. Hierarchical classification problems often have as objective the classification of a

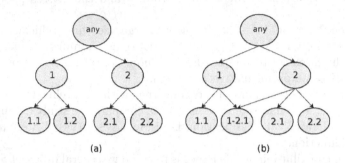

(a)                                        (b)

**Fig. 1.** Examples of hierarchies of classes: (a) structured as a tree and (b) structured as a DAG

new input data into one of the leaf nodes. The deeper the class in the hierarchy, the more specific and useful is its associated knowledge. It may be the case, however, that the classifier does not have the desired reliability to classify a data into deeper classes. In this case, it would be safer to perform a classification into shallower levels of the hierarchy.

In tree structures, the deeper the level, in general the more difficult is the class prediction phase. This may be due to the fact that the classes in deeper levels represent more specific information and are produced by models that have been induced from a smaller number of instances. Therefore, they are more difficult to predict. For DAG structures, the analysis is more complex. As a child node may have more than one parent, some classification models in deeper levels may have been induced from more instances than their ancestral. Besides, in practice, even for DAGs, the prediction accuracy rate decreases with the increase in the class level (depth) [1].

Herewith, the closer the predicted class is to the root of the class tree, the lower the classification error tends to be. On the other hand, such classification becomes less specific and, as a consequence, less useful. Therefore, a hierarchical classifier must deal with the trade-off class specificity versus classification error rate.

In some problems, all instances must be associated to classes in leaf nodes. These problems are named "mandatory leaf-node prediction problems". When this obligation does not hold, the classification problem is a "optional leaf-node prediction problem".

# 3    Classification Approaches for Hierarchical Problems

Following the nomenclature in [1], four types of approaches to deal with these problems may be cited: transformation of the hierarchical problem into a flat classification problem, hierarchical prediction with flat classification algorithms, Top-Down classification and Big-Bang classification.

Several solutions have been proposed for the induction of classification models for hierarchical problems. Many hierarchical classification works have been published in the last years, mainly related to text mining problems [7,6]. Nevertheless, due to the inherent hierarchical characteristic of several biological problems, hierarchical classification has found in the Bioinformatics area a vast and promising exploration field.

Next, a brief description of the four types of hierarchical classification schemes is presented, along with works related to the protein function prediction problem.

## Transformation of the Hierarchical Problem into a Flat Classification Problem

Although in a hierarchical problem the classes are hierarchically organized, this approach reduces the original hierarchical problem to a flat classification problem. This idea is supported by the fact that a flat classification problem may be viewed as a particular case of hierarchical classification, in which there are no subclasses and superclasses. Traditional approaches for flat classification may be applied in this context, without the need to perform alterations or adjustments.

Jensen et al. [8] describe a method, named ProtFun, which uses an ensemble of simple Neural Networks, with a single completely connected intermediate layer, to predict protein categories. The method predicts functional categories as originally defined by Riley [9] for Escherichia coli. The classification model described in Weinert and Lopes [10] is based on Multilayer Perceptron Neural Networks. The model was applied to the classification of functional and structural characteristics of enzymes from the Protein Data Bank [11].

## Hierarchical Prediction with Flat Classification Algorithms

This approach divides a hierarchical problem into a set of flat classification problems. The main difference to the previous approach is the possibility to consider several levels of the hierarchy. In this approach, each class level is treated as an independent classification problem. For each level, flat classification algorithms may then be used.

Clare and King [12] describe a classification method that uses the C4.5 algorithm [13] and was applied to S.cerevisiae phenotype data. Jensen et al. describe in [14] an extension of the method proposed in [8], used in the prediction of GO (Gene Ontology) classes. Laegreid et al. [15] predicts GO classes using a combination of a rule induction algorithm based on the Rough-Set theory [16] and Genetic Algorithms [17]. The method described produces rules that model the relation between the gene expression levels over time in order to predict the bio-logical paper of unknown genes. Tu et al. [18] propose a classification model that uses Neural Networks to infer annotations for more specific classes from known annotations for their superclasses. As application, data from serum response in serum-starved human fibroblasts were used. Barutcuoglu et al. [19] described a method where Support Vector Machines (SVMs) [20] classifiers are trained in-dependently for each class. A Bayesian hierarchical combination scheme is later used to allow error correction collaboration among all nodes.

**Top-Down Hierarchical Classification**

In the Top-Down approach, one or more classifiers are trained for each level of the hierarchy. This produces a tree of classifiers. The root classifier is trained with all training instances. Then, at the next class level, a classifier is training with just the subset of instances belonging to the classes predicted by the classifier. E.g, in the class tree of Fig. 1(a), a classifier associated with the class node "1" would be trained only with instances belonging to class 1.1 or 1.2, but its training set would not include instances of class 2.1 or 2.2. The process of training classifiers proceeds in a top-down fashion until classifiers predicting the leaf class nodes are produced. Hence, the top-down approach follows the well-known "divide-and-conquer" principle.

In the test phase, beginning at the root node, an instance is classified in a Top-Down manner. When assigned to one class, the instance is then submitted to a new classifier in order to predict to which of this class' subclasses it belongs. This procedure is repeated until a leaf-node class is reached or until no addi-tional prediction can be made from an internal node, such that the reliability is not affected. As this approach performs the classification through a modular process, the classifier induction is simpler when compared to the Big-Bang ap-proach, described next. In particular, although it produces a tree of classifiers, each classifier is built by running a flat classification algorithm. Nevertheless, its disadvantage is that errors made in higher levels of the hierarchy are propagated to the more specific levels.

Holden and Freitas [21] proposed a hybrid algorithm that combines characteristics of the PSO (Particle Swarm Optimization) [22] and ACO (Ant Colony Optimization) [23] techniques for the induction of rule-based classifiers in a Top-Down manner. The hybrid algorithm was employed for the classification of enzymes. In [24], Holden and Freitas used the same hybrid algorithm proposed in [21], with some extensions, for the classification of G-Protein-Coupled Receptors [25].

**Big-Bang Hierarchical Classification**

One can consider that truly hierarchical classification algorithms are instances of the Big-Bang and the Top-Down approaches [1]. In the Big-Bang approach, a classification model is created in a single run of the algorithm, considering the hierarchy of classes as a whole, presenting then a higher algorithmic com-plexity. After the classification model training, the prediction of the class of a new instance is carried out in just one step. For this reason, in contrast to the other approaches, Big-Bang cannot use pure flat classification techniques. If a flat classification technique is used in the Big-Bang approach, it must be adapted to consider the whole hierarchy.

Clare and King [26] modified the C4.5 learning algorithm to predict functional classes from S.cerevisiae ORFs (Open Reading Frames). Blockeel et al. [27] used a Decision Tree induction algorithm based on the notion of predictive clustering trees [28]. The algorithm generates as output one tree for the whole hierarchy of classes. Phenotype S.cerevisiae data were used in the experiments.

# 4   Materials and Methods

## 4.1   Datasets

The datasets used in this paper employ signatures (describing sequence similarity) generated directly from protein sequences to attempt to predict a given protein's function. The two datasets used in this paper involve G-Protein-Coupled Receptor (GPCR) and Enzyme protein families.

G-protein-coupled receptors are proteins involved in signalling. They span cell walls so that they influence the chemistry inside the cell by sensing the chemistry outside the cell. More specifically, when a ligand (a substance that binds to a protein) is received by a GPCR, it causes the attached G-proteins to activate and detach. This is a mechanical biological switch that causes the released G-Protein to affect other reactions within the cell. This kind of protein is particularly important for medical applications because it is believed that $40\% - 50\%$ of current medical drugs target GPCR activity [29].

Enzymes are another subset of proteins; they are catalysts which are used to speed up and make possible many of the chemical reactions that take place within the cell, without being altered themselves during the reaction. They are usually very specific, only catalysing one type of reaction within the cell. Often they can be turned on and off by a ligand (a small molecule that interacts with the enzyme). This is used to control both the speed of reaction and the course of overall reaction pathways that take place within the cell.

The protein functional classes are given unique hierarchical indexes by [25] in the case of GPCRs and by Enzyme Commission Codes [3] in the case of enzymes. In the case of GPCRs, proteins (data instances) have up to five class levels, but only four levels are used in the datasets created in this work, as the data in the $5th$ level is too sparse for training - i.e., in general there are too few instances of each class at the $5th$ level. All four levels of the Enzyme Commission Codes are used in the created Enzymes datasets.

The datasets used in our experiments were constructed from data extracted from UniProt [30] and GPCRDB [25]. UniProt is a well known biological database, containing sequence data and a rich annotation about a large number of different kinds of proteins. It also has cross-references for other major biological databases. UniProt was extensively used in this work as a source of data for creating the datasets used in the experiments. Only the UniProtKB/Swiss-Prot was used as a data source, as it contains a higher quality, manually annotated set of proteins. Unlike Uniprot, GPCRDB is a biological database specialised on GPCR proteins.

The predictor attributes in the two datasets are Interpro entries [31,32], along with the molecular weight and sequence length of each protein. Interpro integrates several protein signature databases (Gene3D, PANTHER, PIRSF, Pfam, PRINTS, PROSITE, SMART, SUPERFAMILY and TIGRFAM) giving a very powerful "representation language" to describe the main patterns or "motifs" (e.g., specific sub-sequences of amino acids) present in a given protein or group of proteins. The component protein signature databases from which Interpro entries are derived use three main methods of protein identification: PROSITE uses regular expressions, PRINTS uses groups of non-overlapping motifs and the rest rely on Hidden Markov Model methods.

Any duplicate instances (proteins) in a dataset are removed in a preprocessing step, i.e., before the hierarchical classification algorithm is run, to avoid redundancy. For both GPCR and Enzyme datasets, if there are fewer than ten instances in any given class in the class tree that class is merged with its parent class. If the parent class is the root node, the entire small class is removed from the dataset. This process helps to ensure there is enough instances per class to allow the classifier to perform a reasonably reliable prediction of each class. Any binary attribute that has a value which occurs in only one instance is removed from the corresponding dataset, since these binary attributes in general do not have a good predictive power. An initial random sample of 15000 enzymes from the UniProt database was used to generate the enzyme datasets. Less than the original 15000 instances occur in the final datasets because of the duplicate and small class removal process.

After preprocessing the datasets used in the experiments, the GPCR dataset ended up with 450 predictor attributes, 7461 instances (proteins) and 12/54/82/50 classes per level (number of classes at level 1/2/3/4, respectively). The Enzyme dataset presented 1216 predictor attributes, 14036 instances and 6/41/96/187 classes per level. Due to a high computational cost, the Enzyme dataset was reduced to 6925 instances and 2/21/48/87 classes per level.

Both datasets were divided according to the 5-fold cross-validation method. Accordingly, each dataset is divided into five parts of approximately equal size. At each round, one fold is left for test and the remaining folds are used in the classifiers training. This makes a total of five train and test sets. The final accuracy rate of a classification model is then given by the mean of the predictive accuracy on the test set obtained for each fold.

## 4.2    Decision Trees

A Decision Tree (DT) is a data structure containing two types of nodes, namely: a leaf node that corresponds to a class or a decision node that contains a test over some attribute. For each test result, there is an edge for a subtree. In the classification of a new instance in the DT, the tree is traversed according to the tests' results in a top-down fashion until a leaf node is reached. The instance is labeled with the class associated to this node. Examples of DT induction algorithms are the ID3 algorithm [4] and its successor C4.5 [13]. This work employed the C4.5 algorithm in the DT induction.

Besides being a practical method for concept learning from data [5], DT models have a high comprehensibility, that is, the knowledge acquired by the tree during its training is easy to understand and interpret. These were the motivations for the choice of this particular technique for classifier induction in this work.

### 4.3   Hierarchical Classification Models

The four hierarchical methods described in Section 3 are compared in this work for the protein classification problems investigated. The first considers only the leaf nodes of the problem hierarchy, inducing a flat classifier that distinguishes all classes associated to this set. The idea is that the classification of a new instance in a class associated with a leaf node also implies in its classification in classes at higher (shallower) levels of the tree. E.g, if an instance is classified as 2.1.3.4, then the instance is considered assigned to class 2 at the first level, class 2.1 at the second level, and so on, in order to compute the predictive accuracy per level reported later. The second approach decomposes the hierarchical problem into a set of flat classification problems, each one distinguishing all classes present in a level of the hierarchy. The third method uses the Top-Down approach and the last one, the Big-Bang approach. All of them use DT induction algorithms to produce the classification models. These approaches were chosen in order to compare different schemes for hierarchical classification.

The flat and Top-Down approaches were implemented using the package TREE of the R tool [33]. The Big-Bang approach used was the one developed by Clare and King [26]. This method uses a modified version of the C4.5 algorithm, called HC4.5. The original code of HC4.5 can automatically assign a new instance to a class in any level of the tree, depending on the characteristics of the data at each level. Since the goal of this paper is to do an experiment comparing the Big-Bang and other approaches in a way which is as fair and controlled as possible, we modified HC4.5, including the restriction that it always assigns a new instance to a class in a leaf node of the class tree. This automatically assigns to the instance classes at higher levels of the class tree too.

## 5   Experiments

Experiments were performed in order to evaluate the hierarchical classification methods described in Section 4.3 using the datasets from Section 4.1.

### Results

The mean accuracy results obtained in the GPCR dataset 5-fold cross-validation partitions are shown in Table 1. This table shows, for each level of the GPCR hierarchy, the mean accuracy rates of the hierarchical classifiers induced. The standard deviation rates of the accuracies obtained in the cross-validation data partitions are illustrated in parentheses. The accuracy rate corresponds to the percentage of correctly classified patterns in a dataset.

**Table 1.** Accuracy results in the GPCR dataset

|         | Flat Classif. based on leaves | Flat Classif. all levels | Top-Down     | Big-Bang     |
|---------|-------------------------------|--------------------------|--------------|--------------|
| Level 1 | 61.33 (0.62)                  | 87.80 (0.37)             | 87.80 (0.37) | 91.13 (0.97) |
| Level 2 | 57.11 (0.54)                  | 68.64 (0.43)             | 74.12 (0.65) | 76.05 (1.69) |
| Level 3 | 21.97 (0.29)                  | 29.22 (0.54)             | 46.17 (2.12) | 43.38 (1.01) |
| Level 4 | 31.36 (1.28)                  | 58.17 (2.73)             | 73.60 (4.46) | 68.02 (4.96) |

Like Table 1, Table 2 shows the mean and standard deviation accuracy results observed for the Enzyme dataset partitions.

**Table 2.** Accuracy results in the Enzyme dataset

|         | Flat Classif. based on leaves | Flat Classif. all levels | Top-Down     | Big-Bang     |
|---------|-------------------------------|--------------------------|--------------|--------------|
| Level 1 | 82.73 (1.22)                  | 89.78 (0.85)             | 89.78 (0.85) | 88.97 (0.36) |
| Level 2 | 61.82 (1.03)                  | 60.33 (1.98)             | 73.75 (1.34) | 84.56 (0.84) |
| Level 3 | 58.24 (1.08)                  | 53.79 (2.68)             | 61.38 (1.24) | 84.13 (0.82) |
| Level 4 | 59.17 (1.48)                  | 58.93 (0.66)             | 59.93 (0.13) | 96.36 (0.43) |

## Discussion

The high performance obtained by all approaches in the first level of the EC dataset, shown in Table 2, occurred because the first level of this dataset has only 2 classes, different from the GPCR dataset, which presents 12 classes in the first level.

According to the results showed in tables 1 and 2, the Top Down and Big Bang approaches performed better than the flat approaches for all levels in both datasets. This was expected, once the Top Down and the Big Bang approaches consider the hierarchy during their training and test. This makes the prediction in deeper levels easier. For the flat approaches, the accuracy tends to decrease faster than the hierarchical approaches with the increase of the levels depth.

For the GPCR dataset, the flat approach based on all levels performs significantly better than the flat approach based on the leaf nodes. For the EC dataset, none of the flat approaches is clearly superior to the other.

Regarding the hierarchical approaches, for the GPCR dataset, the Top Down algorithm has a lower accuracy than the Big Bang algorithm for the first two levels and a higher accuracy for the last two levels. For the EC dataset, the Big Bang is clearly better than the Top Down in the last three levels. This difference may be due to the different hierarchical structure and the class (and instances per class) distribution in the hierarchy of these datasets.

Regarding the class distribution, the GCPR dataset has a reduced number of classes, and instances, in the fourth level, when compared with the EC dataset. This occurs because all classes in the fourth level of the GCPR dataset are part of the subtree rooted in the class 1 of the first level. The other classes in the first level have descendents only in the levels 2 and 3. The fourth level has 50 of the total 198 classes. The EC dataset, in opposite, has 87 of the total 158 classes in the fourth level.

The unbalanced nature of the distribution of classes in GPCR dataset seems to favour the correct prediction in the last levels by the Top Down algorithm. A possible reason is the error propagation mechanism employed by this algorithm (see Section 3). Since several leaf nodes are in the intermediate levels of the hierarchy, the errors are not propagated to the deepest levels. Besides, as most of the last level classes are descendents of the class 1, which has the highest correct prediction rate, the propagation of errors to the descendents of this class are less severe.

The GPCR dataset has 1544 of instances in the fourth level, from a total of 7500, making 20.59% of the instances. The EC dataset, on the other hand, has 4887 of instances in the fourth level, from a total of 6995, making 69.86% of the instances. A similar situation occurs in the third level. We believe that the reduced number of classes and instances in the last levels harms the performance of the Big Bang algorithm. This hapens because, unlike the Top Down algorithm, which uses a divide-and-conquer mechanism for the classification in the leaf nodes, the Big Bang predictions are made directly in the leaf nodes. For the previous reason, the high number of instances in the last level of the EC dataset may have favoured the Big Bang algorithm, see Table 2.

## 6   Conclusions

In this paper, we presented a comparative study of hierarchical approaches based on decision trees. Four approaches for hierarchical classification were investigated. Two approaches based on flat classification, the Big Bang approach and the Top Down approach.

In order to evaluate the performance of these approaches, experiments were performed using two bioinformatics datasets, which are related with G-Protein-Coupled Receptor (GPCR) and Enzyme Protein (EC) families. According to the experimental results, the Top Down and the Big Bang approaches performed better than the two flat approaches for all levels in both datasets. In the EC dataset, the Big Bang approach outperformed the Top Down approach in the last 3 levels. In the GPCR dataset, the Top Down approach was clearly superior in the last two levels.

For future work, the authors plan to investigate the performance of the hierarchical approaches when the deepest classification level assigned to each test instance is automatically defined by the system, without the restriction of always assigning one of the leaf classes to every test instance. Other hierarchical classification algorithms will also be investigated. Finally, the authors plan to combine hierarchical classification with multi-label classification.

**Acknowledgments.** The authors would like to thank the Brazilian research councils FAPESP and CNPq for their financial support and Amanda Clare for the Big-Bang code used in the experiments.

# References

1. Freitas, A.A., Carvalho, A.C.P.F.: A Tutorial on Hierarchical Classification with Applications in Bioinformatics. In: Taniar, D. (ed.) Research and Trends in Data Mining Technologies and Applications, Idea Group, pp. 176–209 (2007)
2. Blake, J.: Gene Ontology(GO) Tutorial, [Online; accessed April 07, 2006] (2003), http://www.geneontology.org/teaching_resources/tutorials/2003_MBL_jblake.pdf
3. E. Nomenclature, of the IUPAC-IUB. p. 104, American Elsevier Pub. Co., New York, NY (1972)
4. Quinlan, J.R.: Induction of decision trees. Machine Learning 1(1), 81–106 (1986)
5. Mitchell, T.M.: Machine Learning. McGraw-Hill Higher Education, New York (1997)
6. Sun, A., Lim, E.P., Ng, W.K.: Hierarchical text classification methods and their specification. Cooperative Internet Computing 256, 18 (2003)
7. Sun, A., Lim, E.P.: Hierarchical text classification and evaluation. In: Proceedings of the 2001 IEEE International Conference on Data Mining, pp. 521–528. IEEE Computer Society Press, Washington, DC, USA (2001)
8. Jensen, L.J., Gupta, R., Blom, N., Devos, D., Tamames, J., Kesmir, C., Nielsen, H., Stærfeldt, H.H., Rapacki, K., Workman, C., Andersen, C.A.F., Knudsen, S., Krogh, A., Valencia, A., Brunak, S.: Prediction of human protein function from post-translational modifications and localization features. Journal of Molecular Biology 319(5), 1257–1265 (2002)
9. Riley, M.: Functions of the gene products of Escherichia coli. Microbiology and Molecular Biology Reviews 57(4), 862–952 (1993)
10. Weinert, W.R., Lopes, H.S.: Neural networks for protein classification. Applied Bioinformatics 3(1), 41–48 (2004)
11. Bernstein, F.C., Koetzle, T.F., Williams, G.J., Meyer, E.F., Brice, M.D., Rodgers, J.R., Kennard, O., Shimanouchi, T., Tasumi, M.: The Protein Data Bank. A computer-based archival file for macromolecular structures. FEBS Journal 80(2), 319–324 (1977)
12. Clare, A., King, R.D.: Knowledge Discovery in Multi-label Phenotype Data. In: Siebes, A., De Raedt, L. (eds.) PKDD 2001. LNCS (LNAI), vol. 2168, pp. 42–53. Springer, Heidelberg (2001)
13. Quinlan, J.R.: C4.5: Programs for Machine Learning. Morgan Kaufmann, San Francisco (1993)
14. Jensen, L.J., Gupta, R., Stærfeldt, H.H., Brunak, S.: Prediction of human protein function according to Gene Ontology categories. Bioinformatics 19(5), 635–642 (2003)

15. Laegreid, A., Hvidsten, T.R., Midelfart, H., Komorowski, J., Sandvik, A.K.: Predicting Gene Ontology Biological Process From Temporal Gene Expression Patterns. Genome Research 13(5), 965–979 (2003)
16. Pawlak, Z.: Rough Sets: Theoretical Aspects of Reasoning about Data. Kluwer Academic Publishers, Norwell, MA, USA (1992)
17. Mitchell, M.: An Introduction to Genetic Algorithms. Mit Press, Cambridge (1996)
18. Tu, K., Yu, H., Guo, Z., Li, X.: Learnability-based further prediction of gene functions in Gene Ontology. Genomics 84(6), 922–928 (2004)
19. Barutcuoglu, Z., Schapire, R.E., Troyanskaya, O.G.: Hierarchical multi-label prediction of gene function. Bioinformatics 22(7), 830–836 (2006)
20. Cristianini, N., Shawe-Taylor, J.: An Introduction to Support Vector Machines and other kernel-based learning methods. Cambridge University Press, Cambridge (2000)
21. Holden, N., Freitas, A.A.: A hybrid particle swarm/ant colony algorithm for the classification of hierarchical biological data. In: Proceedings of the 2005 IEEE Swarm Intelligence Symposium, pp. 100–107. IEEE Computer Society Press, Los Alamitos (2005)
22. Sousa, T., Silva, A., Neves, A.: Particle swarm based Data Mining Algorithms for classification tasks. Parallel Computing 30(5-6), 767–783 (2004)
23. Parpinelli, R.S., Lopes, H.S., Freitas, A.A.: Data mining with an ant colony optimization algorithm. IEEE Transactions on Evolutionary Computation 6(4), 321–332 (2002)
24. Holden, N., Freitas, A.A.: Hierarchical Classification of G-Protein-Coupled Receptors with PSO/ACO Algorithm. In: Proceedings of the 2006 IEEE Swarm Intelligence Symposium, pp. 77–84. IEEE Computer Society Press, Los Alamitos (2006)
25. GPCRDB, Information system for G protein-coupled receptors (GPCR), [Online; accessed July 2006] (2006), URLhttp://www.gpcr.org/7tm/
26. Clare, A., King, R.D.: Predicting gene function in Saccharomyces cerevisiae. Bioinformatics 19(90002), 42–49 (2003)
27. Blockeel, H., Bruynooghe, M., Dzeroski, S., Ramon, J., Struyf, J.: Hierarchical multi-classification. In: Proceedings of the ACM SIGKDD 2002 Workshop on Multi-Relational Data Mining (MRDM 2002), pp. 21–35. ACM Press, New York (2002)
28. Blockeel, H., Raedt, L.D., Ramon, J.: Top-down induction of clustering trees. In: Proceedings of the Fifteenth International Conference on Machine Learning, pp. 55–63 (1998)
29. Filmore, D.: It's a GPCR world. Modern drug discovery 1(17), 24–28 (2004)
30. Apweiler, R., Bairoch, A., Wu, C.H., Barker, W.C., Boeckmann, B., Ferro, S., Gasteiger, E., Huang, H., Lopez, R., Magrane, M., et al.: UniProt: the Universal Protein knowledgebase. Nucleic Acids Research 32, D115–D119 (2004)
31. Interpro [Online; accessed July 2006] (2006), http://www.ebi.ac.uk/interpro/
32. McDowall, J.: InterPro: Exploring a Powerful Protein Diagnostic Tool. In: ECCB05, Tutorial, p. 14 (2005)
33. Venables, W.N., Smith, D.M.: The R Development Core Team, An introduction to R - version 2.4.1 (2006), http://cran.r-project.org/doc/manuals/R-intro.pdf

# High Efficiency on Prediction of Translation Initiation Site (TIS) of RefSeq Sequences

Cristiane N. Nobre[1], J. Miguel Ortega[2], and Antônio de Pádua Braga[3]

[1] Bioinformática, UFMG
nobre@pucminas.br
[2] Laboratório de Biodados, ICB, UFMG
miguel@icb.ufmg.br
[3] Engenharia Eletrônica, UFMG
apbraga@cpdee.ufmg.br

**Abstract.** An important task in the area of gene discovery is the correct prediction of the translation initiation site (TIS). The TIS can correspond to the first AUG, but this is not always the case. This task can be modeled as a classification problem between positive (TIS) and negative patterns. Here we have used Support Vector Machine working with data processed by the class balancing method called Smote (Synthetic Minority Over-sampling Technique). Smote was used because the average imbalance has a positive/negative pattern ratio of around 1:28 for the databases used in this work. As a result we have attained accuracy, precision, sensitivity and specificity values of 99% on average.

**Keywords:** Translation Initiation Site, Support Vector Machine, Smote, Imbalanced Data.

## 1 Introduction

Only a portion of the mature messenger RNA (mRNA) is translated into a polypeptide chain. This portion is known as the CDS, or CoDing Sequence, which is flanked by 5'UTR and 3'UTR UnTranslated Regions [1]. Thus, every transcript containing a CDS bears a pattern known as the Translation Initiation Site (TIS). Pedersen and Nielsen [2] call attention to the fact that the TIS does not always cohabit with the first start codon AUG in the mature mRNA. Moreover, it is not even guaranteed that the first ATG in a cDNA sequence will be known, especially if they are searched for in ESTs (Expressed Sequence Tags, single-pass partial cDNA sequences).

This situation can be modeled as a classification problem between positive (TIS) and negative (pseudoTIS) patterns that can be addressed by using the Support Vector Machine (SVM) classifier. However, since only one occurrence of the TIS is expected against several pseudoTIS in a typical mRNA sequence (the average imbalance is a ratio of around 1:28 between positive:negative patterns for the databases used in this work), any analysis will require a method for class balancing. One such method is the Synthetic Minority Over-sampling Technique (Smote)[3].

In a seminal work, Stormo and colleagues [4] studied the TIS in *E. coli* using perceptron [5] and windows of 51, 71 or 104 bases and binary codification (A=1000,

M.-F. Sagot and M.E.M.T. Walter (Eds.): BSB 2007, LNBI 4643, pp. 138–148, 2007.

C=0100, G=0010 e T=0001), and characterized the Shine Dalgarno pattern in 124 genes.

The first initiative devoted to eukaryotic genes was conducted in 1984 by Kozak [6] who determined a consensus for a large collection of data. The Kozak consensus is GCC[AG]CCatgG, where "G" is frequent in +4 position and a purine, preferably an "A", is present in position -3. Frequently, eukaryotic ribosome scans the mRNA until it reaches the first AUG from the 5' region [6][7][8] and starts translation. However, this is not a rule, and a subsequent AUG codon merged into a more effective pattern is used. Translation can also start in a different codon from AUG, although this does not often happen in eukaryotic genes [8][9].

Pedersen e Nielsen [2] used an Artificial Neural Network (NN) trained using a database composed of Genbank [10] sequences from vertebrates and reached an efficiency of 85%. Sequences were filtered to remove introns and redundancy was reduced by eliminating excess representatives of gene families and homologous genes. In addition to this, only sequences bearing at least 10 upstream or 150 downstream bases from the annotated start codon were selected. This database has subsequently been used by others, the result being a database of 13503 sequences containing 3312 TIS (24.5%) and 10190 pseudoTIS (75.5%). However, gene annotation has progressed and a database of reference sequences (RefSeq) is now available [11].

Using SVM, Zien and colleagues [12] improved the analysis of Pedersen e Nielsen's database using the same window size and binary codification. The improvement was attributed to modifications in the kernel, reaching an accuracy level of 88.1%. Later, with another kernel modification (Salzberg Kernel), an accuracy level of 88.6% was attained for the same database.

Hatzigeorgiou [9] reported the development of the software DIANA-TIS which uses two Neural Networks and complete human sequences of cDNA to reach an efficiency level of 94%. In this implementation, ribosome scanning is important, since a linear search starts in 5'UTR and stops when a positive score is found. Therefore, it was assumed that complete sequences would be used as input.

Zeng and colleagues [13], using 100 bases flanking the TIS and a concept of the analysis of characteristics, have reached 90% accuracy using Pedersen e Nielsen's dataset. Adding Hatzigeorgiou's scanning approach [9] they increased the accuracy to 94.4%.

Huiqing and colleagues [14] used $k$-gram amino acid patterns, top-ranked features and a classification SVM model or ensembles of decision trees to predict the TIS, improving the results of Zeng et al. [13].

Haifeng and Tao [15] introduced a class of new sequence-similarity kernels based on string edit, edit kernels, for use with SVMs. Moreover, they converted the region of an input mRNA downstream sequence into an amino acid sequence before applying SVMs. Accuracy, sensitivity and specificity were 99.9%, 99.92% and 99.82%, respectively, using 30 upstream bases and 180 downstream bases.

Tzanis and colleagues[16][17][18] also rely on characteristics to detect TIS and reach an accuracy level of 96.25% extracting them from a window of 99 upstream nucleotides and 99 downstream nucleotides from ATG.

Here we evaluate the use of Smote on the improvement of the SVM classifier. Moreover, we concentrate the analysis on a small window and we use the reviewed

*Mus musculus* and *Rattus norvegicus* RefSeq sequences downloaded from the National Center for Biotechnology Information as the analysis data. As a result we have attained accuracy, precision, sensitivity and specificity values of 99% on average.

## 2 Methods

Our methodology consists of five phases: (1) the choice of the database that should be analyzed; (2) the choice of the codification to be used; (3) the execution of the data balancing; (4) the choice of the classifier that offers a good performance in this classification; and, finally, (5) the choice of measurements of performance to be used.

### 2.1 Database

Sequences from *Mus musculus* and *Rattus norvegicus* were downloaded from *RefSeq* NCBI ftp site. Entries not containing at least 12 upstream bases from the TIS were discarded. From the six levels of confidence available (*reviewed, provisional, predicted, validated, model and inferred*) only the *reviewed* sequences were used. Negative patterns were extracted registering whether they were in frame or out of frame with respect to the TIS, or if they were upstream or downstream TIS (cf Fig. 1).

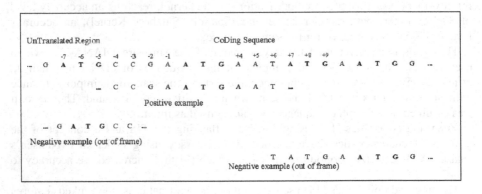

**Fig. 1.** Construction of the positive and negative patterns using 24 nucleotide windows (12 nucleotides in the upstream region and 12 in the downstream region) with an ATG codon starting at the 13th position

**Table 1.** Negatives Patterns extracted from RefSeq sequences. Of the negative examples in the training, only those that are out of frame were considered.

| Organism | Position and in (IF) or out (OF) of frame | | | |
|---|---|---|---|---|
| | 5'IF | 5'OF | 3'IF | 3'OF |
| *Mus musculus* | 96 | 221 | 3990 | 6943 |
| *Rattus norvegicus* | 15 | 35 | 485 | 900 |

A dataset of 182 and 42 positive patterns (TIS) and 7164 and 935 negative patterns (pseudoTIS out of frame) were obtained from *Mus musculus* and *Rattus norvegicus*, respectively, distributed according to position and frame as shown in table 1.

## 2.2  Binary Codification

Bases were codified as: A=0001, C=0010, G=0100 and T=1000. Outputs were set equal to 1 for the TIS and 0 for pseudoTIS (negative pattern). A complementary codification was created which used joint bases (tri-nucleotides) instead of individual ones: AAA=000000, AAC=000001, AAG=000002, AAT=000003 and so on. Tri-nucleotides were positioned in frame to the candidate ATG. In other words, instead of codifying base by base, the codification was done by tri-nucleotides with the intention of reducing the number of inputs to the SVM and consequently reducing the processing time.

## 2.3  Class Balance

The Smote algorithm described by Nitesh [3] was used for replication of examples from the minority class. In other words, examples of the minority class were interpolated between examples of the sample, resulting in classes with an equivalent number of patterns. The dataset used here showed an initial imbalance in the average proportion of 1:28 for TIS:pseudoTIS patterns.

## 2.4  Support Vector Machine

Characterized as a machine learning algorithm capable of resolving linear and non-linear classification problems, the main idea of classification by support vector is to separate examples with a linear decision surface and to maximize the margin of separation between the other training points [19][20][21][22].

The SVM works as follows: Given a set of training data $\{x_i, y_i\}_{i=1}^{N}$, each with an input vector $x_i \in \Re^n$ and corresponding binary output $y_i \in \{-1, +1\}$, the objective is to separate the class -1 vectors from the class +1 vectors.

The SVM *light* version implemented by T. Joachims [23] and available at http://svmlight.joachims.org was used here. A 4th order polynomial function was adopted.

## 2.5  Efficiency

Four parameters were applied to measure efficiency and these are represented by the formulas below. We evaluated accuracy, precision, sensitivity and specificity.

$$accuracy = \frac{TP + TN}{TP + TN + FN + FP} \qquad (1)$$

$$precision = \frac{TP}{TP + FP} \qquad (2)$$

$$sensitivity = \frac{TP}{TP + FN} \qquad (3)$$

$$specificity = \frac{TN}{TN + FP} \qquad (4)$$

where, TP = true positive, TN = true negative, FP = false positive and FN = false negative. Thus, accuracy is the proportion of all predictions that are correct; precision is the proportion of the apparent TIS that are indeed TIS; sensitivity refers to the proportion of TIS correctly classified as TIS, while specificity is the proportion of pseudoTIS correctly classified as pseudoTIS.

### 2.6 Validation

All results were obtained using 5-fold cross validation [24]. In other words, the dataset was divided into five groups. Four groups were used for training and the fifth one was reserved for testing. The procedure was repeated for the four training groups and results were averaged, the standard deviation then being calculated.

## 3 Results and Discussion

### 3.1 Window Size

Two distinct windows sizes were chosen for comparison, both of them symmetrically flanking the TIS. Results are shown in table 2. Additional variations have been tested (data not shown) but the 12+12 combination was shown to yield results that are similar to larger windows.

Results indicate that a sequence of 12 upstream bases and 12 downstream bases of the TIS is sufficient to obtain a good level of efficiency. This is remarkably important for sequences where few bases surrounding the putative TIS are

**Table 2.** Efficiency as a function of window size flanking the TIS

|  | windows 12 + 12 | | windows 99 + 99 | |
| --- | --- | --- | --- | --- |
| | *Mus musculus* | | | |
| | Mean | Std deviation | Mean | Std deviation |
| Accuracy | 98.79 | 0.26 | 98.83 | 0.60 |
| Precision | 97.69 | 0.54 | 97.81 | 1.14 |
| Sensitivity | 99.95 | 0.07 | 99.90 | 0.12 |
| Specificity | 99.32 | 0.12 | 98.83 | 0.60 |
| | *Rattus norvegicus* | | | |
| | Mean | Std deviation | Mean | Std deviation |
| Accuracy | 99.73 | 0.14 | 98.38 | 0.80 |
| Precision | 99.58 | 0.24 | 99.02 | 1.60 |
| Sensitivity | 99.88 | 0.17 | 99.31 | 0.20 |
| Specificity | 99.73 | 0.14 | 98.86 | 0.02 |

available. Moreover, as the number of entries for the classifier increases linearly with the number of bases used, processing time is optimized. For example, in the *Mus musculus* base, the processing time is 7.3 lower if we use 12+12 sized windows. This can be of fundamental importance if we are working with much larger bases than this.

## 3.2  Codification

Table 3 presents a comparison between two options for codification. While binary codification is the most commonly used in the literature, we proposed a codification joining all three bases (tri-nucleotides) flanking the TIS, similar to a codon. Note that this procedure is not used as a sliding window.

**Table 3.** Efficiency obtained as a function of the codification used

| | Codification by tri-nucleotides | | Codification by base | |
|---|---|---|---|---|
| | *Mus musculus* | | | |
| | Mean | Std deviation | Mean | Std deviation |
| Accuracy | 98.79 | 0.26 | 99.99 | 0.01 |
| Precision | 97.69 | 0.54 | 98.58 | 0.24 |
| Sensitivity | 99.95 | 0.07 | 99.88 | 0.11 |
| Specificity | 99.32 | 0.12 | 99.02 | 0.14 |
| | *Rattus norvegicus* | | | |
| | Mean | Std deviation | Mean | Std deviation |
| Accuracy | 99.73 | 0.14 | 97.23 | 0.14 |
| Precision | 99.58 | 0.24 | 98.32 | 0.54 |
| Sensitivity | 99.88 | 0.17 | 98.89 | 0.20 |
| Specificity | 99.73 | 0.14 | 99.20 | 0.16 |

These results show that the codification proposed here yielded accuracy levels close to those obtained with the standard codification using individual bases. It is noteworthy that the new codification reduces the number of entries by half, resulting in a reduction of the size of the input file that could be compatible with implementation in most regular desktop computers. It should be remembered that with the *Mus musculus* base, the processing time is 19.09 lower if we use codification by tri-nucleotides. This gain, added to the gain from the window size, can make a great difference in much larger bases than the ones used in this project.

## 3.3  Selection of PseudoTIS Patterns Used for Training

Considering that the balance in favor of pseudoTIS patterns is high, a possible way to decrease the contamination of pseudoTIS with some positive patterns is to discard those pseudoTIS that are in frame with the TIS. In reality, some proteins may have two in frame TIS. Results shown in table 4 support the *elimination* of in frame pseudoTIS from the training step.

**Table 4.** Efficiency as a function of selection of pseudoTIS patterns used for training

| | In frame | | Out of frame | |
|---|---|---|---|---|
| | **Mus musculus** | | | |
| | **Mean** | **Std deviation** | **Mean** | **Std deviation** |
| Accuracy | 90.88 | 3.79 | 98.79 | 0.26 |
| Precision | 84.75 | 7.34 | 97.69 | 0.54 |
| Sensitivity | 96.92 | 2.74 | 99.95 | 0.07 |
| Specificity | 74.25 | 10.60 | 99.32 | 0.12 |
| | **Rattus norvegicus** | | | |
| | **Mean** | **Std deviation** | **Mean** | **Std deviation** |
| Accuracy | 92.88 | 5.02 | 99.73 | 0.14 |
| Precision | 82.34 | 7.06 | 99.58 | 0.24 |
| Sensitivity | 97.26 | 1.03 | 99.88 | 0.17 |
| Specificity | 72.05 | 12.90 | 99.73 | 0.14 |

The efficiency of the classifier is clearly reduced by using negative patterns that are in the same reading frame as the TIS . Additionally, we have analyzed all out of frame pseudoTIS with both 5' out of frame pseudoTIS and 3' out of frame pseudoTIS, and concluded that the best results are obtained using all *out of frame* pseudoTIS, regardless of their position with respect to the TIS (upstream or downstream).

## 3.4   Class Balance

All results described have made use of Smote. The importance of this procedure is illustrated by the data shown in table 5. Without class balancing, efficiency is greatly reduced.

**Table 5.** Efficiency as a function of balancing method

| | Without Smote | | With Smote | |
|---|---|---|---|---|
| | **Mus musculus** | | | |
| | **Mean** | **Std deviation** | **Mean** | **Std deviation** |
| Accuracy | 98.50 | 0.24 | 98.79 | 0.26 |
| Precision | 82.01 | 9.24 | 97.69 | 0.54 |
| Sensitivity | 51.11 | 11.48 | 99.95 | 0.07 |
| Specificity | 93.60 | 2.60 | 99.32 | 0.12 |
| | **Rattus norvegicus** | | | |
| | **Mean** | **Std deviation** | **Mean** | **Std deviation** |
| Accuracy | 96.45 | 0.23 | 99.73 | 0.14 |
| Precision | 90.30 | 1.35 | 99.58 | 0.24 |
| Sensitivity | 56.43 | 15.56 | 99.88 | 0.17 |
| Specificity | 95.30 | 1.68 | 99.73 | 0.14 |

Remarkably, sensitivity increases from 51% to 99.95% and from 56% to 99.88% in *M. musculus* and *R. norvegicus, respectively.*

## 3.5  Scanning Model

It is desirable to know the frequency of all pseudo TIS classified as negative and the TIS classified as positive for the entire population of molecules. Moreover, by random chance a negative pattern may simulate a functional TIS, although this might be a rare event, and such patterns are likely to occur downstream of the TIS. Thus, only one positive prediction should occur per molecule during a scan that starts from the 5' extremity. This prediction schedule was introduced by Agarwal e Bafna [25] and later used by Hatzigeorgiou [9] and others [13] [14]. To address this model, we examined whether or not a given molecule possesses only negative patterns in the upstream region of the actual TIS, and, furthermore, if it was predicted as positive. This index was calculated for *Mus musculus* and *Rattus norvegicus*, respectively, as 95.87 +/- 1.23 percent and 98.54 +/- 1.89 percent, using the proposed conditions (12+12 window, codification by tri-nucleotides, out of frame pseudoTIS, Smote). These settings yielded even better results than the other variations (data not shown).

## 3.6  False Positive TIS

An analysis of the frequency of each base flanking TIS suggests that false positives selected by the above method (fig. 2c), although not entirely reproducing the pattern shown for true positives (fig. 2a), do not approach the random distribution that is peculiar to true negatives (fig. 2b). Fig. 2a shows the typical Kozak consensus which obeys the rules for position -3 (purine) and +4 (guanine) and which has a high frequency of G in position -6. Clearly, as seen for the true positive, false positive patterns do not demonstrate a random distribution pattern for the frequency of bases. This shows that the classifier ranks the false positive patterns separately from the negative ones.

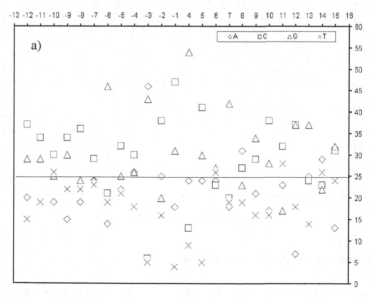

**Fig. 2.** Analysis of the frequency of (a) positives, (b) negatives and (c) false-positive sequences, for each position

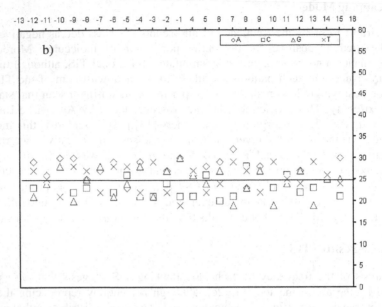

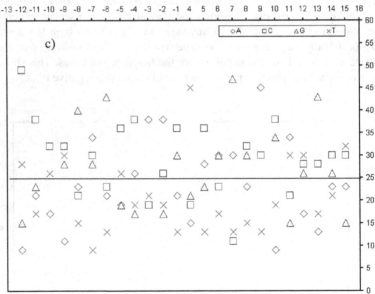

**Fig. 2.** (*continued*)

# 4   Perspectives and Future Work

The SVM classifier used under the conditions evaluated here shows remarkable advantages over other similar implementations. Binary codification by tri-nucleotides decreased the size of the input file and the time of analysis. The use of out of frame negative patterns is a simple implementation that provides good results. Class balancing has played an important role in the improvement of the training step and Smote has adapted well to the problem of TIS prediction. Some issues remain to be addressed and are being investigated. Is it possible to train the classifier with sequences from a given organism to analyze cDNA sequences from a new organism? Furthermore, what is the minimum number of sequences needed to reduce the standard deviation of the predictions to an acceptable level? Are gene predictors used to generate RefSeq prediction sequences accurate? Now that TIS prediction with SVM has been set up under the parameters described here, these questions, amongst others, can now be asked and answered.

## Acknowledgements

The authors are thankful to A.G. Pedersen and G. Tzanis for sharing important data and to Saulo Paula Pinto for critically reviewing the manuscript.

## References

1. Zien, A., Ratsch, G., Mika, S., Scholkopf, B., Lemmen, C., Smola, A., Lengauer, T., Muller, K.R.: Engineering support vector machine kernels that recognize translation initiation sites. In: Proc. German Conference on Bioinformatics '99, pp. 37–43 (1999)
2. Pedersen, A.G., Nielsen, H.: Neural network prediction of translation initiation sites in eukaryotes: perspectives for EST and genome analysis. In: Proc. 5th International Conference on Intelligent Systems for Molecular Biology, pp. 226–233 (1997)
3. Chawla, N.V., Bowyer, K.W., Hall, L.O., Kegelmeyer, W.P.: SMOTE: Synthetic Minority Over-sampling Technique. Journal of Artificial Intelligence and Research 16, 321–357 (2002), Disponível em citeseer.ist.psu.edu/chawla02smote.html
4. Stormo, G.D., Schneider, T.D., Gold, L.M.: Characterization of translational Initiation sites. E. coli. Nucleic Acid Res. 10, 2971–2996 (1982)
5. Haykin, Simon: Redes Neurais: princípios e prática. Bookman (2001)
6. Kozak, M.: Compilation and analysis of sequences upstream from the translational start site in eukaryotic mRNAs. Nucleic. Acids Research 12, 857–872 (1984)
7. Kozak, M.: An analysis of 5'-noncoding sequences from 699 vertebrate messenger RNAs. Nucleic. Acids Research 15, 8125–8148 (1987)
8. Kozak, M.: The scanning model for translation: an update. J. Cell. Biol. 108, 229–241 (1989)
9. Hatzigeorgiou, A.G.: Translation initiation start prediction in human cDNAs with high accuracy. Bioinformatics 18, 343–350 (2002)
10. Benson, D., Boguski, M., Lipman, D., Ostell, J.: Genbank. Nucleic Acids Research. 25, 1–6 (1997)
11. Pruitt, K.D., Maglott, D.R.: Refseq and locuslink: NCBI Gene-centered resources. Nucleic Acids Research. 29, 137–140 (2001)
12. Zien, A., Ratsch, G., Mika, S., Scholkopf, B., Lemmen, C., Smola, A., Lengauer, T., Muller, K.-R.: Engineering support vector machine kernels that recognize translation Initiation sites. Bioinformatics 16, 799–807 (2000)

13. Zeng, F., Yap, R.H., Wong, L.: Using feature generation and feature selection for accurate prediction of translation initiation sites. Genome Informatics Ser Workshop Genome Informatics 13, 192–200 (2002)
14. Liu, H., Han, H., Li, J., Wong, L.: Using amino acid patterns to accurately predict translation initiation sites. Silico Biology 4, 0022 (2004)
15. Li, H., Jiang, T.: A class of edit kernels for SVMs to predict translation initiation sites in eukaryotic mRNAs. In: Proceedings of the Eighth International Conference on Research in Computational Molecular Biology, San Diego, California, USA, pp. 262–271 (2004)
16. Tzanis, G., Berberidis, C., Alexandridou, A., Vlahavas, I.: Improving the Accuracy of Classifiers for the Prediction of Translation Initiation Sites in Genomic Sequences. In: Bozanis, P., Houstis, E.N. (eds.) PCI 2005. LNCS, vol. 3746, pp. 11–13. Springer, Heidelberg (2005)
17. Tzanis, G., Berberidis, C., Vlahavas, I.: A Novel Data Mining Approach for the Accurate Prediction of Translation Initiation Sites. In: Maglaveras, N., Chouvarda, I., Koutkias, V., Brause, R. (eds.) ISBMDA 2006. LNCS (LNBI), vol. 4345, pp. 92–103. Springer, Heidelberg (2006)
18. Tzanis, G., Vlahavas, I.: Prediction of Translation Initiation Sites Using Classifier Selection. In: Antoniou, G., Potamias, G., Spyropoulos, C., Plexousakis, D. (eds.) SETN 2006. LNCS (LNAI), vol. 3955, pp. 367–377. Springer, Heidelberg (2006)
19. Carvalho, B.P.R., Almeida, M.B., Braga, A.P.: Support Vector Machines - um estudo sobre técnicas de treinamento. Technical Report Monogra_a interna no.3, Universidade Federal de Minas Gerais, Belo Horizonte, MG (2002)
20. Boser, B.E., Guyon, I., Vapnik, V.: A training algorithm for optimal margin classifiers. In: Computational Learing Theory, pp. 144–152 (1992)
21. Burbidge, R., Buxton, B.: An introduction to support vector machines for data mining. In: M. Sheppee (Ed.) Keynote Papers, Young OR12, University of Nottingham, 3.15 Operational Research Society: Operational Research Society (2001)
22. Scholkopf, B., Mika, S., Burges, C.J.C., Knirsch, P., Muller, K.R., Ratsch, G., Smola, A.J.: Input space versus feature space in kernel-based methods. IEEE Transactions on Neural Networks 10, 1000–1017 (1999)
23. Joachims, T.: Making large-Scale SVM Learning Practical. In: Schölkopf, B., Burges, C., Smola, A. (eds.) Advances in Kernel Methods - Support Vector Learning, MIT-Press, Cambridge (1999), http://www-ai.cs.uni-dortmund.de/DOKUMENTE/joachims_99a.pdf
24. Kohavi, R.: A Study of Cross-Validation and Bootstrap for Accuracy Estimation and Model Selection. In: Proceedings of 14th International Joint Conference on Artificial Intelligence (IJCAI) (1995)
25. Agarwal, P., Bafna, V.: The ribosome scanning model for translation initiation for gene prediction and full-length cDNA detection. In: Proc. 5th International Conference on Intelligent Systems for Molecular Biology, pp. 2–7 (1998)

# Outlining a Strategy for Screening Non-coding RNAs on a Transcriptome Through Support Vector Machines

Roberto T. Arrial[1], Roberto C. Togawa[2], and Marcelo de M. Brígido[1]

[1] Departamento de Biologia Celular, Universidade de Brasília
70910-900, Brasília, DF, Brazil
[2] Laboratório de Bioinformática, EMBRAPA-Cenargen
70770-900, Brasília, DF, Brazil
rtarrial@gmail.com

**Abstract.** Evidences that non-coding RNAs exert functions in organisms accumulate in the literature. Both computational predictions and experimental results have shown that, albeit not coding for a protein product, these transcripts play roles as diverse as catalytic activities and complex gene regulations, suggesting its therapeutic potential when applied to the study of pathogenic organisms. A target for such approach is the fungus *Paracoccidioides brasiliensis* (Pb), the ethyological agent of paracoccidioidomycosis, whose transcriptome has recently been elucidated. This work reports the compiling of a large training set and implementation of a framework of programs for sequence feature extraction, generating input for a Support Vector Machines algorithm for characterizing the coding potential of transcripts from a transcriptome.

**Keywords:** non-coding RNA; ncRNA; *Paracoccidioides brasiliensis*; transcriptome; Support Vector Machines; machine learning.

## 1 Introduction

Transcriptome sequencing projects generate a wealth of information, most of which must be analyzed and interpreted in the light of bioinformatics studies. Accordingly, the quality and quantity of data to be extracted from these projects are limited by the current theoretical models and paradigms, and by the repertoire of available computational tools for analysis.

For a long time the main focus of these projects was the mRNA, carriers of the information necessary for coding proteins. It was not until recently that the focus also included non-coding RNAs (ncRNAs), a class of functional molecules present in all Kingdoms of life, with varied activities such as protein transport, gene regulation, silencing, imprinting, among others [1]. Although the ncRNA biological model quality has evolved considerably, the ncRNA study is neither an easy task, nor is it well established, regardless of the approach. Computational methods are impaired by the absence of obvious intrinsic or extrinsic ncRNA signals, which could possibly allow for reliable labeling of a transcript as non-coding, and also the need for

M.-F. Sagot and M.E.M.T. Walter (Eds.): BSB 2007, LNBI 4643, pp. 149–152, 2007.

appropriate tools, exploiting not only sequence comparison but also secondary structure motifs and other important RNA properties.

Despite the lack of standard tools for identifying ncRNAs (like BLAST stands for mRNAs), perhaps the most successful computational field in the task of screening for new ncRNA is the machine learning, specially the Support Vector Machines (SVMs). This algorithm uses features from known examples to derivate a model for predicting characteristics of unknown data, unseen during the training phase [2]. This strategy has already been successfully employed for the prediction of several ncRNA species, such as siRNAs, miRNAs and snoRNAs. This work discusses the adaptation of the method to the problem of identifying specifically the ncRNA mRNA-like, which are transcripts of variable sizes, stabilities and functions, commonly found mixed with mRNAs on transcriptome analyzes. To this end, features extracted from a training set containing known examples of both ncRNA and mRNA are provided for a SVM program for later predicting, in a transcriptome context, which transcripts are most likely coding and non-coding.

The present work reports the initial steps for constructing the algorithm, comprising the assembly of the training set, its pre-processing and the formulation of the attributes to be extracted. As a benchmarking procedure, the trained algorithm will be used to classify transcripts from the *Paracoccidioides brasiliensis* (Pb) fungus [3]. Once optimized, there is interest in making the algorithm available for analyzes of transcriptomes from both eukaryotic and prokaryotic organisms, as there were no phylogenetic filters imposed during training set construction.

## 2 Materials and Methods

### 2.1 Training Set Construction

The training set consists of a positive set, comprising known mRNA (protein-coding) sequences, and a negative set, which contains known ncRNA (non-protein-coding) sequences. The positive set was built as follows. A file with the Swiss-Prot database version 50.8 was downloaded in October, 2006. The FASTA file containing 234,112 protein sequences had its redundant sequences (>70% similarity in length) eliminated using the CD-HIT program. PERL scripts were used to parse the Swiss-Prot IDs from the remaining sequences, which in turn were used to recover its corresponding cDNA ID from the nucleotide database EMBL. The cDNA IDs were then submitted to the EBI dbFetch service, which allow retrieving sequences. The resulting nucleotide database is again submitted to elimination of redundancy through the program BLASTCLUST, and the remaining sequences have their ORFs determined by the program ORFPredictor. Then, the positive set is split in two: dbCOD_N, consisting of transcripts with predicted ORFs, and dbCOD_P, comprising the predicted proteins.

The negative set was built in a similar way. Files containing sequences from the NONCODE, RNAdb and Rfam databases (version October, 2006) were downloaded, joined and formatted. The FASTA file containing 265,691 ncRNA sequences had its redundancy eliminated through BLASTCLUST, and the remaining sequences had its ORFs predicted using ORFPredictor. The nucleotide sequences comprised the dbNC_N set, and its predicted protein products, dbNC_P.

The training set construction is summarized in Figure 1.

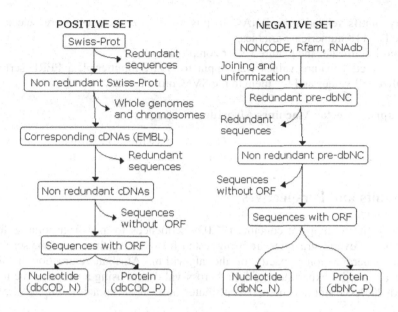

**Fig. 1.** Training set construction scheme. Vertical arrows denote sequential steps; horizontal pointing arrows indicate discarding of specified elements.

## 2.2 Feature Vector

A feature vector is the assembly of several attributes, arbitrarily selected by the researcher, which describe the data in some way.

**Table 1.** Selected features, corresponding feature extraction programs and amount of variables allocated to each attribute. Qualitative features are shown already quantified.

| Feature description | Program | Variables |
|---|---|---|
| 1. Nucleotide composition | SVM_inputter.pl | 84 |
| 2. ORF length | SVM-inputter.pl | 4 |
| 3. Amino acid composition | SVM-inputter.pl | 20 |
| 4. Protein isoelectric point | iep (EMBOSS) | 1 |
| 5. Protein complexity | CAST | 1 |
| 6. Protein intrinsic unfolding | FoldIndex | 1 |
| 7. Protein functional domains | ProSite Scan | 1 |
| 8. Known homolog proteins | BLAST | 1 |
| 9. Mean protein hydropathy | SVM-inputter.pl | 1 |
| 10.Protein secondary structure | SSPro 4.0 | 3 |
| 11.Protein solvent accessibility | ACCPro 4.0 | 1 |

In this work, the features where chosen based on their value to represent some aspects from a typical non-coding transcript or from a real protein. It is expected that a putative translation from a ncRNA carries many unlikely properties, which, when combined and analyzed together, may allow for its identification as a "false" protein. On the other hand, features extracted from the transcript sequence may help to identify motifs related to ncRNAs. In this work, 11 attributes were selected and encoded as 118 numerical variables.

Table 1 delineates the feature vector composition.

All extracted feature values were input to SVM_inputter.pl, a PERL script that normalizes all values and fits them to the SVM input format.

### 2.3 Support Vector Machines Algorithm Settings

LIBSVM v2.84 was chosen, set as C-SVM (classification problem), binary (two-classes), with the RBF Kernel. This setting is broadly adopted in similar problems.

## 3 Results and Perspectives

Currently, the training set contains 126,039 sequences formed from joining dbCOD and dbNC. Several strategies are being tested for reducing the size of the set without loss in generalization capacity of the algorithm. Attribute extraction is already underway. A pilot experiment of 5-fold cross-validation using a subset of the training set with 20,000 instances and only 9 attributes indicates accuracies of up to 91.5%.

**Acknowledgments.** This work was supported by CNPq, National Counsel of Technological and Scientific Development - Brazil.

## References

1. Mattick, J.S., Makunin, I.V.: Non-coding RNA. Hum. Mol. Genet. 15, R17–R29 (2006)
2. Noble, W.S.: What is a support vector machine? Nat. Biotech. 24, 1565–1567 (2006)
3. Felipe, M.S., et al.: Transcriptional profiles of the human pathogenic fungus Paracoccidioides brasiliensis in mycelium and yeast cells. JBC 280, 24706–24714 (2005)

# Mapping Contigs onto Reference Genomes*

Nalvo F. Almeida**, André C. Lima, Said S. Adi,
Carlos J.M. Viana, Marcel Y. Nakazaki, Andrey A. Tamura,
Luciana Y. Hiratsuka, and Leandro P. Brazil

Department of Computing and Statistics,
Federal University of Mato Grosso do Sul
CP 549, 79070-900, Campo Grande, MS, Brazil
nalvo@dct.ufms.br
http://egg.dct.ufms.br/projects

**Abstract.** This work presents a preliminary comparative study of some
tools for mapping annotated contigs onto a close-related complete annotated genome. This kind of mapping could help scientists in generating
additional sequences to fill in gaps in finishing genome projects, or even
in getting relevant functional information, specially when annotations of
the complete genome and contigs are available.

**Keywords:** comparative genomics, mapping contigs.

## 1 Introduction

Although nowadays genome sequencing is fast and cheap, one of the most difficult
steps in a genome project is closing the gaps. In an usual whole genome shotgun
project, the assembly results in hundreds or thousands of contigs. On the other
hand, when the complete genome of some very related organism has become
available, it is possible, by comparing all the contigs to the complete genome,
to determine the order and orientation of the contigs of the incomplete genome,
besides the approximated distances among them. This mapping of contigs onto
a reference genome could help scientists in generating additional sequences to fill
in the gaps, or even in getting a good set of functional information, enough to
finish off the project without finishing the genome, specially when annotations
of the complete genome and contigs are available.

In this work we present a preliminary comparative study of four tools for
mapping annotated contigs onto a complete annotated genome: NUCmer [7],
PROmer [7], Mega BLAST [13] and EGG [1]. The need for this came from some
attempts to align the contigs of a brazilian isolated of the bacteria *Anaplasma
marginale* [3], which is being sequenced in state of Mato Grosso do Sul, Brazil,
with the complete genome of *Anaplasma marginale* str. St. Maries [4]. This
organism is the causative agent of bovine anaplasmosis and is responsible for

---

* This work was supported by CNPq and Fundect-MS.
** Corresponding author.

M.-F. Sagot and M.E.M.T. Walter (Eds.): BSB 2007, LNBI 4643, pp. 153–157, 2007.

serious economic damage. Our attempts in mapping those contigs using simple
BLAST programs [2] and MUMmer [6], which is the core tool used for NUcmer
and PROmer, showed some significant discrepancies.

## 2   Methodology

MUMmer [6] is a well-known system to align whole genome sequences based on
suffix trees, an efficient data structure to find all distinct subsequences in a given
sequence. MUMmer has variants for comparing incomplete genomes: NUCmer
and PROmer [7], both using MUMmer as their core alignment engines. NUCmer
is a multiple-contig alignment program. It takes as input two multi-fasta files
representing partial or complete assemblies and returns an alignment of every
sequence contig in the first multi-fasta file to every sequence in the second one.
PROmer translates the DNA to amino acids before comparing the sequences.

Mega BLAST [13] makes use of an optimized greedy algorithm for nucleotide
sequence alignment search, aiming mainly to align sequences that differ slightly.
Mega BLAST is also able to efficiently handle much longer DNA sequences than
the traditional BLAST programs.

EGG (Extended Genome-Genome comparison) [1] makes whole genome pairwise comparison by finding all pairs of orthologous genes using BLASTP program [2]. The comparison takes all the predicted proteins of both genomes as
input, following an all-against-all fashion and build a bipartite graph, where each
edge represents a pair of orthologous genes, called a *match*. Formally, a match is
a pair $(g, h)$ of genes whose BLASTP e-values (both ways) are not greater than
$10^{-5}$ and the alignments include at least 60% of each sequence. When a gene $h$
is the best BLASTP hit found by $g$ and vice versa, we have a *bi-directional best
hit (BBH)*. Thus, a gene can participate at most of one BBH. When a gene $g$
found no BLASTP hits on the other genome, we say that $g$ is a *specific* gene.
EGG has been used successfully in several genome projects [5,8,10,12].

Any strategy of mapping contigs onto a genome is useful only for pairs of very
close-related species and even for those cases it is almost impossible to know *a
priori* what is the best mapping, due the existence of genome rearrangements.
We have built our methodology for comparing the tools by fragmenting a whole
chromosome in non-overlapping contigs and mapping them onto the same chromosome. This strategy is based on the assumption that any mapping method
should be able to correctly map the contigs onto the genome where they came
from. Table 1 shows the ten genomes we have used in our analysis.

We define a score $S_M$ to assess the quality of a mapping $M$ as follows. Given a
chromosome $G$ with coordinates $[1 \ldots n]$ and $C = \{C_1 \ldots C_k\}$ the set of $k$ contigs
of $G$, let $P_i$ be the first base of contig $C_i$ in $G$. Let $X_i$ be the position where
the method $M$ mapped the first base of $C_i$ in $G$, $1 \leq i \leq k$, called *mapping of*
$C_i$. We call *distance of mapping* of $C_i$ in $G$, and denote it by $d_{M,i}$, the distance
between $P_i$ and $X_i$, according to the coordinates and both strands of $G$ and
the mapping built by $M$. Thus, $d_{M,i} = |P_i - X_i|$. In case of $G$ is a circular
chromosome, $d_{M,i} = \min\{|P_i - X_i|, n - |P_i - X_i|\}$. The score $S_M$ is defined as

**Table 1.** Each genome has been randomly broken in 100 or 1000 non-overlapping contigs, randomly distributed in both strands

| Genome | Genome size | 100 Contigs | | 1000 Contigs | |
|---|---|---|---|---|---|
| | | Mean size of contigs | Mean number of genes | Mean size of contigs | Mean number of genes |
| *Bacillus anthracis* Steme | 5227293 | 27283.82 | 26.8 | 2937.83 | 2.14 |
| *Buchera aphidicola* str. Sg | 416380 | 3396.83 | 2.15 | 353.86 | 0.02 |
| *Candidatus blochmannia* str. BPEN | 705557 | 4277.46 | 2.48 | 439.11 | 0.03 |
| *Escherichia coli* W3110 | 4639675 | 26832.18 | 23.56 | 2575.6 | 1.58 |
| *Haemophilus Influenzae* 86 | 1830138 | 10575.84 | 9.08 | 1076.34 | 0.37 |
| *Helicobacter pylori* HPAG1 | 1667867 | 8972.54 | 7.81 | 893.61 | 0.25 |
| *Streptococcus pneumoniae* R6 | 2046115 | 10178.93 | 9.24 | 1122.82 | 0.46 |
| *Synechococcus* sp. CC9605 | 2606748 | 12824.09 | 12.77 | 1352.16 | 0.7 |
| *Xanthomonas campestris* 33913 | 5175554 | 28263.98 | 22.02 | 2807.84 | 1.49 |
| *Xanthomonas campestris* 8004 | 5076188 | 27197.52 | 21.19 | 2853.62 | 1.6 |

the number of distances of mapping $d_{M,i}$ such that $d_{M,i} \leq \alpha \cdot |C_i|$, where $\alpha$ is a factor that allows approximated solutions (not so far from $P_i$). We do not consider, in computing $S_M$, the contigs mapped onto the wrong strand and the ones not mapped at all.

EGG maps proteins of each contig onto the reference genome, namely BBHs, instead of mapping whole contigs. Thus, we need to translate a set of BBHs formed by genes of $C_i$ and genes of $G$ into $X_i$. Formally, let $(g, h)$ be a BBH, where $g$ is gene of $G$ and $h$ is gene of $C_i$. We simulate a Cartesian coordinate system where $G$ and $C_i$ are two horizontal axes and the $(x, y)$-coordinates of $g$ and $h$ are given by $(x_g, 1)$ and $(x_h, 0)$. $X_i$ is calculated by finding a position $\delta$ in $G$ that minimizes the summation of euclidian distances between all BBH-pairs of genes of $G$ and $C_i$, considering $C_i$ shifted $\delta - 1$ positions in $G$. Thus, $X_i$ is given by the formula below and Figure 1 illustrates two BBHs and the shift $\delta$.

$$X_i = \arg \min_{\delta=1...n} \sum_{\text{all BBHs in } C_i \times G} \sqrt{1 + (x_g - x_h - \delta)^2}$$

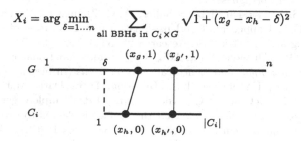

**Fig. 1.** An illustration of two BBHs, $(g, h)$ and $(g', h')$, mapped onto $G$

## 3    Results and Discussion

Taking as input the set of ten genomes of Table 1, randomly broken into 100 and 1000 contigs, we run the programs considering several values for $\alpha$. Table 2 shows the results for $\alpha = 0.10$ and $0.50$. The main evidences from a preliminary analysis showed us that the problem is easier for larger contigs. For small contigs, EGG

**Table 2.** Mean values of $S_M$ for the genomes of table 1, considering $\alpha = 0.5$ and 0.1

| Tools | 1000 contigs | | 100 contigs | |
|---|---|---|---|---|
| | 0.5 | 0.1 | 0.5 | 0.1 |
| Mega Blast | 948.4 | 769.9 | 84.5 | 60.0 |
| EGG | 399.9 | 399.9 | 95.0 | 95.0 |
| NUCmer | 993.5 | 993.5 | 100.0 | 100.0 |
| PROMER | 995.5 | 995.5 | 100.0 | 100.0 |

could not correctly map all of them, since it is based in gene mapping, and those genes could not be present in the contigs. NUCmer had a better performance with larger contigs. On the other hand, for larger contigs Mega BLAST presented bad results, because small local aligments may occur also in other locations of the reference genome. Finally, PROMER presented the best results in all situations, suggesting that translation of the contigs may help the whole mapping.

This is an initial analysis and much remains to be done. One next step is to improve the time-consuming algorithm for mapping contigs from BBHs. Another step is to include misleading bases and also repeats into the contigs, in order to simulate more real situations in the mapping problem. Also, the building of a new mapping tool, including the use of scaffolds [9] when they are available, may be promising, since scaffolds may come with a partial order of the contigs and approximated distances among them. This analysis is part of a bigger project that includes, among others, additional analyses and some tools for comparing genomes (http://egg.dct.ufms.br/projects). This web site also presents graphical display of some mappings using GBrowse [11] environment.

# References

1. Almeida, N.F.: Tools for genome comparison. PhD thesis, IC-Unicamp, Campinas-SP, Brazil (In Portuguese) (2002)
2. Altschul, S.F., Madden, T.L., Schaffer, A.A., Zhang, J., Zhang, Z., Miller, W., Lipman, D.J.: Gapped blast and psi-blast: a new generation of protein database search programs. Nucleic Acid Research 25, 3389–3402 (1997)
3. Araújo, F.R., Almeida, N.F., Adi, S.S., et al.: Functional characterization of *Anaplasma marginale* and identification of antigens (In preparation)
4. Brayton, K.A., Kappmeyer, L.S., Herndon, D.R., Dark, M.J., Tibbals, D.L., Palmer, G.H., McGuire, T.C., Knowles Jr., D.P.: Complete genome sequencing of *Anaplasma marginale* reveals that the surface is skewed to two superfamilies of outer membrane proteins. Proc. Natl. Acad. Sci. USA 102(3), 844–849 (2005)
5. da Silva, A.C.R., Setubal, J.C., Almeida, N.F., et al.: Comparison of the genomes of two *Xanthomonas* pathogens with differing host specificities. Nature 417(6887), 459–463 (2002)
6. Delcher, A.L., Kasif, S., Fleischmann, R.D., Peterson, J., White, O., Salzberg, S.L.: Alignment of whole genomes. Nucleic Acids Research 27(11), 2369–2376 (1999)
7. Delcher, A.L., Phillippy, A., Carlton, J., Salzberg, S.L.: Fast algorithms for large-scale genome alignment and comparison. NAR 30(11), 2478–2483 (2002)
8. Monteiro-Vitorello, C.B., Camargo, L.E.A., Setubal, J.C., Almeida, N.F., et al.: The genome sequence of the gram-positive sugarcane pathogen *Leifsonia xyli* subsp. *xyli*. Molelular Plant-Microbe Interactions 17(8), 827–836 (2004)

9. Setubal, J.C., Wernec, R.: A program for building contig scaffolds in double-barrelled shotgun genome sequencing. Technical report, IC, Unicamp (2001)
10. Van Sluys, M.A., Setubal, J.C., Almeida, N.F., et al.: Comparative analyses of the complete genome sequences of pierce's disease and citrus variegated chlorosis strains of *Xylella fastidiosa*. J. Bacteriology 185(3), 1018–1026 (2003)
11. Stein, L.D., Mungall, C., et al.: The generic genome browser: a building block for a model organism system database. Genome Res. 12(10), 1599–1610 (2002)
12. Wood, D.W., Setubal, J.C., Almeida, N.F., et al.: The genome of the natural genetic engineer *Agrobacterium tumefaciens* c58. Science 294, 2317–2323 (2001)
13. Zhang, Z., Schwartz, S., Wagner, L., Miller, W.: A greedy algorithm for aligning dna sequences. J. Comput. Biol. 7(1-2), 203–214 (2000)

# Molecular Dynamics Simulations of Cruzipains 1 and 2 at Different Temperatures

Priscila V. S. Z. Capriles and Laurent E. Dardenne

Grupo de Modelagem Molecular de Sistemas Biológicos,
Laboratório Nacional de Computação Científica, LNCC/MCT,
Av. Getúlio Vargas, 333 - 25.651-075 Petrópolis,
Rio de Janeiro, RJ - Brasil
{capriles,dardenne}@lncc.br

**Abstract.** Nearly 100 years after the discovery of *Trypanosoma cruzi*, the parasitic agent of Chagas' disease, there are no appropriate therapies that lead to cure the acute or the chronic phases of this disease. Among the enzymes of *T. cruzi*, already considered as molecular targets for Chagas' disease treatment, the cysteine proteases had been extensively studied by experimental approaches. In the present work, the isoforms 1 and 2 of cruzipain were investigated by molecular dynamics simulations (MD) at 25°C and 37°C temperatures, using as control papain, the representative enzyme of cysteine proteases family C1. The main results showed that the presence of a negatively charged amino acid at the 158 position (papain numbering) in the catalytic site, could induces a structural reorganisation, susceptible to temperature variations, in the catalytic residues CYS25 and HIS159.

**Keywords:** Molecular Dynamics Simulations (MD), *Trypanosoma cruzi*, Cruzipains and Cysteine Proteases.

The protozoa *Trypanosoma cruzi*, parasitic agent of Chagas' disease, is endemic in South and Central America and in Mexico [1], with some related cases in Canada, United States and Europe [2]. The Chagas' disease affects approximately 18 million people with 21,000 dies per year, and at least 100 million persons are exposed to the risk of infection [3].

Nearly 100 years after the discovery, there are no appropriate therapies that lead to cure this disease in the acute or the chronic phases. The incidence, the death rates, the drugs toxicity, linked to the parasite ability to develop drug resistance[4], reinforces the importance of developing new chemotherapy against Chagas' disease. Studies of physiologic and biochemical properties from *T. cruzi*, have shown some enzymes as potential molecular targets for the development of new drugs. In drug design studies, the investigation of the dynamical behaviour of an enzyme and its active site can contribute significantly to the development of a new inhibitor [5].

Among the enzymes of *T. cruzi*, already considered as molecular target for Chagas' disease treatment, the cysteine proteases had been extensively studied by experimental approaches. The lysosomal enzyme cruzipain, is the major cysteine protease of *T. cruzi*, whose main functions are the cellular differentiation of the

M.-F. Sagot and M.E.M.T. Walter (Eds.): BSB 2007, LNBI 4643, pp. 158–162, 2007.

parasite and the host cell invasion. The analysis of electrostatic properties in the catalytic site of cruzipain, showed that its catalytic activity is linked to the existence of a very particular electrostatic environment, tha is responsible for the formation and stabilisation of the ionic pair $CYS25^- \cdots HIS162^+$, as showed in papain [6]. It has been proposed that the catalytic activity of cruzipain would be modulated by structural alterations in the catalytic site, as well as by the presence of a possible allosteric site. The regulation of this allosteric site would be conditioned to the temperature, being inhibited by excess of the substrate at 25°C (temperature found in the invertebrate vector), disappearing at 37°C (temperature found in the vertebrate host) [7].

In the present work, the isoforms $1^1$ and $2^2$ of cruzipain were investigated by molecular dynamics simulations (MD), at 25°C and 37°C temperatures, using as control papain[3], the representative enzyme of cysteine proteases family C1.

The 10 ns of MD analysis (after 1.5 ns of equilibration) were performed by the GROMACS program (version 3.2.1) in double precision, using the GROMACS force field [11]. The systems of 9PAP, 1ME4 and 1ME4m were immersed in a water cubic boxes, considering a solvatation shell with at least $10\text{Å}$ in each dimension of the macromolecule, using the SPC (Simple Point Charge) water model, and they were neutralised adding, $10\,Cl^-$, $11\,Na^+$ and $2\,Na^+$, respectively. For the treatment of the Coulomb potential, it was used the Reaction Field method with a $R_c = 16\text{Å}$ cutoff and a $\varepsilon_{rf} = 54$ dielectric constant. For the Lennard-Jones potential it was used a $R_c = 14\text{Å}$ cutoff.

The analysis of RMSD (Root Mean Square Deviation) and RMSF (Root Mean Square Fluctuation) showed that the secondary structure and the backbone organisation remain stable along all the simulations for papain and cruzipains 1 and 2. These results do not corroborate the hypothesis of the existence of an allosteric site. However, the monitoring of distances and positions of the side chains of the catalytic residues CYS25, HIS159 and ASN175 and the residue in position 158 (papain numbering), showed important alterations in the structural organisation of the catalytic site (Fig. 1).

In the papain, the imidazole ring of HIS159 suffered a "bend", when simulated at 37°C. Probably, this occurred due the lost of the hydrogen bond (HB) between HIS159:NE2 and ASN175:OD1, favouring the electrostatic interaction between HIS159:NE2 and ASP158:OD2, with the formation of a new HB. This conformational alteration favours the entrance of one water molecule in the catalytic site, establishing a HB with ASN175:OD1.

In cruzipain 1, this "bend" phenomenon occurred in both temperatures simulated. At 37°C, it was also observed a 180° rotation of the HIS162 imidazole ring, with the formation of the hydrogen bonds CYS25:SG $\cdots$ HIS162:NE2 and HIS162: ND1 $\cdots$ ASP161:OD2.

---

[1] Crystallographic Structure of Cruzipain 1 - PDB: 1ME4 ($1.20\text{Å}$ of resolution) [8].

[2] Structural model of cruzipain 2 (1ME4m): obtained by comparative modelling (by Modeller program - version 8.0): (i) sequence target - GenBank: M90067 [9]; (ii) template sequence - the crystal structure of cruzipain 1 (PDB: 1ME4).

[3] Crystallographic Structure of Papain - PDB: 9PAP ($1.65\text{Å}$ of resolution) [10].

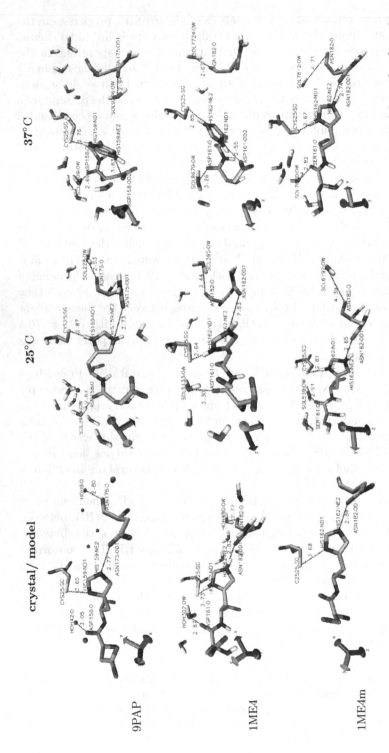

† Treatment of the charge of CYS25: CB= −0.2 and SG= −0.6. These representative conformations were obtained from the clustering method (g_cluster.d function in GROMACS program - version 3.2.1) [11] using as cut-off an RMSD= 1.75Å. The water molecules were selected with criterion of 5.0 Å of distance to ASP158:O/ASP161:O, ASN175:O/ASN182:O and ASN175:OD1/ASN182:OD1.

**Fig. 1.** Distance between the residues from catalytic site of papain (9PAP) and cruzipain 1 (1ME4) e 2 (1ME4m), in the $10^{th}$ ns of molecular dynamic simulation, considering the biological charge of CYS25†

The structure of the ionic pair in cruzipain 2 remained stable along the 10 ns of the MD simulation, in both temperatures. It is important notice that in cruzipain2 occurs the substitution of the negatively charged amino acid ASP161 by a neutral one SER161, what would provide a stabilisation of the usual HB interactions CYS25:SG· · ·HIS162:ND1 and HIS162:NE2· · ·ASN182:OD1.

The MD studies presented in this work, showed that the presence of an acidic residue in position 158 (papain numbering) of the catalytic site, can induces a structural reorganisation, susceptible to temperature variations, of the catalytic amino acids in cysteine proteases from the papain family. This structural reorganisation generates a conformation similar to the one found in serine proteases with the catalytic triad SER−HIS−ASP, and also similar to a serine proteases with the mutation SER→CYS, as presented in the PDB: 1GNS [12].

**Acknowledgements.** The Brazilian National Council of Research (CNPq)and the FAPERJ Foundation have supported this work. Contract grants no. E26/171.199/ 2003, E26/171.401/01, E26/170.648/2004 and CNPq/IM-INOFAR 420.015/ 2005-1.

# References

1. Kirchhoff, L.V., Paredes, P., Lomeli-Guerrero, A., Paredes-Espinoza, M., Ron-Guerrero, C.S., Delgado-Mejia, M., Pena-Munoz, J.G: Transfusion-associated chagas disease (american trypanosomiasis) in mexico: implications for transfusion medicine in the united states. Transfusion 46(2), 298–304 (2006)
2. Reesink, H.W.: European strategies against the parasite transfusion risk. Transfusion Clinique et Biologique 12, 1–4 (2005)
3. WHO, Expert, Committee: Control of chagas disease. Technical Report 905, World Health Organization/ Geneva (2002)
4. Gelb, M.H., Hol, W.G.J.: Drugs to combat tropical protozoan parasites. Science 297(19), 343–344 (2002)
5. Galperin, M.Y., Koonin, E.V.: Searching for drug targets in microbial genomes. Current Opinion in Biotechnology 10, 571–578 (1999)
6. Dardenne, L.E., Werneck, A.S., Neto, M.O., Bisch, P.M.: Eletrostatic properties in the catalytic site of papaisn: a possible regulatory mechanism for the reactivity of the ion pair. PROTEINS: Structure, Funciton and Genetics 52, 236–253 (2003)
7. Lima, A.P., Scharfstein, J., Storer, A.C., Menard, R.: Temperature-dependet substrate inhibiton of the cysteine proteinase (gp57/51) from *trypanosoma cruzi*. Molecular & Biochemical Parasitology 56, 335–338 (1992)
8. Huang, L., Brinen, L.S., Ellman, J.A.: Crystal structures of reversible ketone-based inhibitors of the cysteine protease cruzain. Bioorganic & Medicinal Chemistry 11, 21–29 (2003)
9. Lima, A.P., Tessier, D.C., Thomas, D.Y., Scharfstein, J., Storer, A.C., Vernet, T.: Identification of new cysteine protease gene isoforms in *trypanosoma cruzi*. Molecular & Biochemical Parasitology 67(2), 333–338 (1994)
10. Kamphuis, I.G., Kalk, K.H., Swarte, M.B.A., Drenth, J.: Structure of papain refined at 1.65 Å resolution. Journal of Molecular Biology 179(2), 233–256 (1984)

11. Lindahl, E., Hess, B., van der Spoel, D.: Gromacs 3.0: A package for molecular simulation and trajectory analysis. Journal of Molecular Modeling 7, 306–317 (2001)
12. Almog, O., Gallagher, D.T., Ladner, J.E., Strausberg, S., Alexander, P., Bryan, P., Gilliland, G.L.: Structural basis of thermostability. The Journal of Biological Chemistry 277(30), 27553–27558 (2002)

# Genetic Algorithm for Finding Multiple Low Energy Conformations of Poly Alanine Sequences Under an Atomistic Protein Model

Fábio L. Custódio, Hélio J. C. Barbosa, and Laurent E. Dardenne

Laboratório Nacional de Computação Científica, Av. Getúlio Vargas, 333, 25651-070,
Petrópolis, RJ, Brazil
{flc,hcbm,dardenne}@lncc.br

**Abstract.** The determination of the three-dimensional structure of a
protein is one of the most challenging problems of modern science. A genetic algorithm (GA) was developed to find low energy conformations under an atomist protein model. A crowding method was used for parental
replacement. The comparison criterion between individuals was the absolute RMSD of the $C_\beta$ positions' of the residues. The GROMOS96 force
field potential energy function was used to evaluate the energy of the conformations. We tested the performance of the GA against poly-alanine
sequences of lengths 18 and 23 in a situation where the global minimum
was an alpha helix, and also when it was some other compact structure.
The GA proved very efficient by having a 100% success ratio in finding both the global minimum and the alpha helix conformation in all
situations.

**Keywords:** genetic algorithm, multiple minima, protein structure prediction.

## 1  Introduction

The function of a protein is determined by its structure, and to be able to predict
the native structure of a protein would help unleash the potential of the large
amount of biological sequences information that is being generated by genome
projects. Furthermore, the determination of the three-dimensional structure of
a protein from first principles is one of the most challenging problems of modern
science.

From the point of view of the development of effective optimization algorithms
the protein structure prediction problem is associated with two closely related
crucial aspects: (i) the problem involves thousands of degrees of freedom and
correspond to the exploration of highly complex energy surfaces, and (ii) from
the physical point of view, the ideal cost function involves approaching the system at the atomic level and determining Gibbs' free energy, which requires the
computation of the system's entropy as a whole (molecular system + solvent).
Its determination, even approximately, leads to extremely costly computations
using molecular dynamics techniques. The current impossibility of treating both

M.-F. Sagot and M.E.M.T. Walter (Eds.): BSB 2007, LNBI 4643, pp. 163–166, 2007.
© Springer-Verlag Berlin Heidelberg 2007

aspects above in a computationally feasible way leads to the introduction of simplifications in order to reduce the number of degrees of freedom of the system, and specially to reduce the complexity of the cost function adopted. From the practical point of view, one is dealing with an extremely complex problem using a cost function that frequently does not guarantee that the biologically relevant structure (experimentally found using X-ray diffraction in crystals or NMR techniques) is the global minimum of the hypersurface being investigated. Atomistic models usually rely on empirical potential functions, composing a force fields, which models the many different interactions between the atoms of the protein and the solvent. There is a reasonable degree of uncertainty and it is common that a given function, together with its parameters, is only efficient in describing a particular set of macromolecules under specific circumstances. In this work we used the GROMOS version 96 [1] force field. The fitness function has important approximations (*e.g.* absence of entropic and solvent effects) which translate in doubts about the biological relevance of the lowest energy conformation. From that comes the importance of using a search methodology capable of simultaneously exploring multiple minima on the energy hypersurface to, at the end of a single run, obtain not only the global minimum, but other local minima as well.

## 2    Objectives

Our objectives were to develop a genetic algorithm [2] capable of finding multiple low energy structures for an atomic protein model, and access the performance of the GA using poly-alanine sequences as test cases.

## 3    Methodology

Poly-alanine sequences of lengths 18 and 23 were tested and the results were analyzed from the statistics of 30 independent runs. Those lengths were chosen because they are at the required range for a poly-alanine to fold to a stable helical structure [3]. Using only alaline residues in this first approach has the advantage of reducing complexity as one does not need to change residues' side chains conformations. A maximum of 500,000 function evaluations were allowed with a population of 200 individuals. Two cases of the fitness function were used: (i) a dielectric constant of two (non-polar solvent) with the carboxyl and amine terminals neutralized; (ii) a sigmoidal dielectric function with charged terminals.

### 3.1    The Genetic Algorithm

Candidate structures were encoded in a vector containing the backbone dihedral angles ($\phi$, $\psi$ and $\omega$). The operators used were: two-point crossover, multiple-point crossover operator, random mutation, incremental mutation, compensating mutation and contiguous torsion angles exchange. The crossover points were chosen at random as were the mutation's positions. At the beginning of each run all the operators had the same application probability (the sum was always normalized to 100%). An adaptive scheme was used to derive new probabilities based on operator's performance [4]. All operators were applied to the dihedral angles, but the energy function was evaluated using Cartesian coordinates. The list of torsion angles (internal coordinates) was translated to Cartesian coordinates using constant bonds distances and angles. Multiple Solutions The energy landscape of the model studied has a large number of low energy basins. Some basins lead to a native state energy (structures with biological significance), but others do not (energy traps). Nevertheless different basins may exhibit structures with similar energies, but with different structural motifs. An algorithm able to simultaneously explore multiple basins is particularly useful on such complex models, especially because there are doubts about the energy function. A crowding population replacement strategy was used to maintain good diversity on the population and, at the same time, find multiple low energy solutions. Newly generated conformations competed with the most similar in the previous population. The similarity criteria used was the root-mean-squared-deviation of the internal distances between the beta carbons of all residues.

## 4   Results and Discussion

The GA exhibited the same performance on the sequences of length 18 and 23. Using the dielectric constant of two, the coulomb potential is calculated as if the

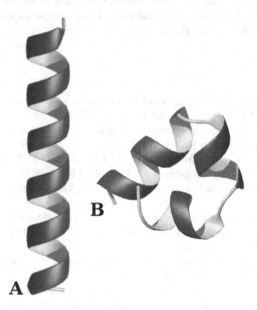

**Fig. 1.** (A) 23-ALA Alpha helix configuration: lowest energy conformation using a dielectric constant of two with the carboxyl and amine terminals neutralized. (B) Lowest energy conformation of 23-ALA with charged carboxyl and amine terminals using a sigmoidal dielectric function.

protein is inside a non-polar environment; this has the effect of exaggerating the electrostatic interactions. Thus it is necessary to neutralize the charged peptide terminals. Under these circumstances, the global energy minimum is an alpha helix structure [5] (Fig.1, A). The GA was able to find with a 100% success rate the alpha helix conformation.

When using a sigmoidal dielectric function [6] and the charged terminals, the global energy minimum is a compact conformation (Fig.1, B) with alpha helix parts and the terminals near to each other. This provides an excellent test situation to assess the ability of the GA in finding structures with biological relevance when they are not the global energy minimum. In all of the runs, the final population contained several structures similar to the global minimum (RMSD < 2.0Å) and a series of structures in alpha helix conformation. These helixes structures had an energy about 9 Kcal/mol higher than the compact conformations.

The presented GA was at first developed for the HP model [7,8] and we adapted it to the atomist model. The algorithm was efficient in finding low energy structures for poly-alanine. The five residue difference on the length did not affect the performance of the algorithm. The crowding parental replacement allowed the GA to simultaneously explore multiple minima, avoiding premature convergence caused by local minima traps, and allowed the GA to find the biological relevant structure even when it was not the global minimum.

**Acknowledgments.** The Brazilian National Council of Research (CNPq) and the FAPERJ Foundation have supported this work. Contract grants no. E26/171.199/2003, E26/171.401/01, E26/170.648/2004.

# References

1. van Gunsteren, W., et al.: Biomolecular Simulation: The gromos96 Manual und User Guide. vdf, Hochschulverlag an der ETH; BIOMOS (1996)
2. Goldberg, D.E.: Genetic Algorithms in Search, Optimization and Machine Learning. Addison-Wesley Longman Publishing Co., Inc., Boston, MA, USA (1989)
3. Moret, M.A., Bisch, P.M., Mundim, K.C., Pascutti, P.G.: New stochastic strategy to analyze helix folding. Biophys J 82(3), 1123–1132 (2002)
4. Davis, L., et al.: Handbook of genetic algorithms. Van Nostrand Reinhold New York (1991)
5. Agostini, F.P., Soares-Pinto, D.D.O., Moret, M.A., Osthoff, C., Pascutti, P.G.: Generalized simulated annealing applied to protein folding studies. J Comput Chem 27(11), 1142–1155 (2006)
6. Arora, N., Jayaram, B.: Strength of hydrogen bonds in $\alpha$ helices. Journal of computational chemistry 18(9), 1245–1252 (1997)
7. Lau, K., Dill, K.: A lattice statistical mechanics model of the conformational and sequence spaces of proteins. Macromolecules 22(10), 3986–3997 (1989)
8. Custódio, F., Barbosa, H., Dardenne, L.: Investigation of the three-dimensional lattice hp protein folding model using a genetic algorithm. Genetics and Molecular Biology 27, 611–615 (2004)

# Cellular Fingerprints: A Novel Concept for the Integration of Experimental Data and Compound-Target-Pathway Relations

## (Extended Abstract)

Stefan Gunther, Stefanie Neumann, Jessica Ahmed, and Robert Preissner

Structural Bioinformatics Group, Institute of Molecular Biology and Bioinformatics, Charité,
Universitätsmedizin Berlin, Arnimallee 22, 14195 Berlin, Germany
http://bioinformatics.charite.de/

## Extended Abstract

The pharmaceutical industry is hunting for high-affinity inhibitors of medical targets, but most of them fail in clinical trials because of severe side effects. On the other hand, there is a growing knowledge about multiple targets and their role in various signalling pathways. Therefore, the integration of experimental data, literature knowledge about drugs, targets, their metabolism, ontology, and related pathways is an important task to achieve better understanding of drug mechanisms on a systems biological level.

To this end we have compiled a database allowing complex queries that project various types of information onto 2,500 WHO-classified drugs [1]. Currently, the database contains about 3,000 target proteins that are annotated by more than 8,000 literature-based and manually curated drug-target relations. The use of the Anatomical-Therapeutical-Chemical drug classification (ATC-code) [2] enables easy access to medical indications or diseases and links phenotypic data to biological processes on a molecular level. Integrated Gene Ontology (GO) [3] enables filtering of proteins associated to particular molecular functions or cellular components. Similarity searching for drugs or targets is implemented via structural fingerprints [4] and FASTA-alignments. Structural fingerprints are bit-vectors encoding for the chemical and topological features of drugs and drug-like compounds. Their similarity can be described using the Tanimoto-coefficient ($T$), considering the concordant and unequal bits of two structural fingerprints [5].

$$T = \frac{N_{ab}}{N_a + N_b - N_{ab}}.$$ (1)

$N_a$ is the number of bits set to 1 in compound $a$, $N_b$ is the number of bits set to 1 in compound $b$ and $N_{ab}$ is the number of bits common to both, compound $a$ and $b$.

For a more comprehensive approach, the public data from the National Cancer Institute on expression and cell effects related to 50,000 compounds are a valuable resource [6]. Besides the expression of all 60 cell types in their basic state, data on the changed

M.-F. Sagot and M.E.M.T. Walter (Eds.): BSB 2007, LNBI 4643, pp. 167–170, 2007.
© Springer-Verlag Berlin Heidelberg 2007

expression after application of cancer drugs with known mechanism of action are deposited. Moreover, data on mutations and properties of the cell panel of the NCI exist [7], which will be useful for the modelling of differences between the cell lines. However, NCI-compounds, which show low cell type specificity were excluded. The remaining data were normalised by z-normalisation to achieve a comparable level of data and to emphasize the differences in the cell specificity. In a next step, the effect on 60 cell lines was translated for each compound into a bit-vector of length 4,800. Thus, this cellular fingerprint describes the unique pattern of effects of a chemical compound on the NCI60 cell panel comparable to the structural fingerprint describing the chemical properties of a compound. Since the average number of cell lines that respond specifically to a single compound is relative low compared to the non-specific cell reactions (about 5/60), an asymmetric distance is an appropriate measure. Again, for fast similarity searching, the Tanimoto-coefficient was used.

The exciting question is whether we will find a correlation between structural and cellular similarity. If the structural similarity between two compounds correlates well with the profile of targets addressed by these compounds, this should be the case. On the other hand, it is known that even for compounds with a distinct structural similarity (Tanimoto > 85%) only one third exhibits similar effects in the same experimental assay [8]. To answer the question for correlation between structural and cellular similarity, a clustering of the compounds with similar cellular effects regarding their structural similarity was performed (Fig. 1). In general, the vast majority of compounds occurs in two or three large clusters, where each cluster represents one structural scaffold. Preliminary analysis shows that one cluster of compounds can be found addressing the same target (Fig. 2) or compounds that address another target in the same pathway. This shows that the cellular fingerprint is indicative of target specificity without structural similarity bias. Examples for coincidence of structural and cellular fingerprints are presented in Fig. 3.

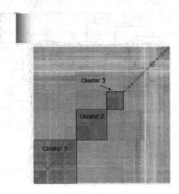

**Fig. 1.** Clustering of compounds with similar cellular fingerprint according to their structural similarity

The presented method supports the following goals:

- Prediction of targets and mechanism of action
  Similarity of structural fingerprints allows hypotheses about similar cellular fingerprints and vice versa. Experimental data on a sufficient number of cell lines allow conclusions about the mechanism of the compound.

- Selection of sensitive cell lines
  Known compound profiles (cellular fingerprints) allow the selection of sensitive cell lines for *in vitro* testing of similar compounds.
- Improved *in silico* screening
  Cellular fingerprints will improve the *in silico* similarity screening. The combination of both methods enables the identification of 'scaffold hoppers', compounds with deviating structure but similar cellular effect.

To validate the correlation between structural and cellular fingerprints experimentally, the purchasable natural compounds from the SuperNatural Database [9] are suitable candidates. Numerous of them are also described in the NCI-database. It will be important to validate predicted cellular fingerprints and target proteins experimentally.

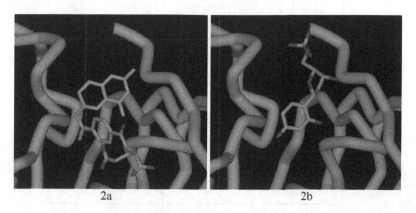

**Fig. 2.** Compounds with different scaffold exhibit similar cellular fingerprints (Methotrexate, 2a and Deoxyuridine phosphate 2b). Both compounds bind to the same target protein (Thymidilate synthase, green; PDB-code: 1AXW).

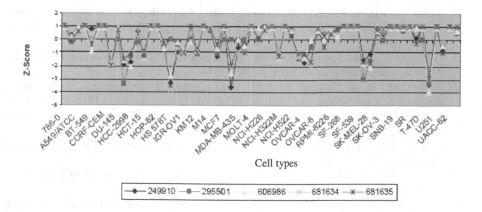

**Fig. 3.** Four compounds with 97% similarity to 7-Chlorocamptothecine and their effect on the NCI60 cell panel. Compare Table 1.

**Table 1.** Results of an *in silico* screening based on cellular fingerprints; the seed of the cluster is 7-chlorocamptothecin (NCS-ID 249910)

| NCS-ID | Cell. Tanimoto | Struct. Tanimoto | Structure |
|---|---|---|---|
| 249910 | 100 | 100 | |
| 681634 | 98.8 | 87.5 | |
| 606986 | 98.3 | 87.7 | |
| 295501 | 98.2 | 96.2 | |
| 681635 | 98.2 | 86.4 | |

# References

1. Goede, A., et al.: SuperDrug: a conformational drug database. Bioinformatics 21(9), 1751–1753 (2005)
2. Skrbo, A., Begovic, B., Skrbo, S.: Classification of drugs using the ATC system (Anatomic, Therapeutic, Chemical Classification) and the latest changes. Med Arh 58(1 Suppl 2), 138–141 (2004)
3. Harris, M.A., et al.: The Gene Ontology (GO) database and informatics resource. Nucleic Acids Res 32(Database issue), D258–D261 (2004)
4. Thimm, M., et al.: Comparison of 2D similarity and 3D superposition. Application to searching a conformational drug database. J Chem Inf Comput Sci 44(5), 1816–1822 (2004)
5. Willett, P.: Similarity-based virtual screening using 2D fingerprints. Drug Discov Today 11(23-24), 1046–1053 (2006)
6. Collins, M.J.: The NCI Developmental Therapeutics Program. Clin Adv Hemtol Oncol 4(4), 271–273 (2006)
7. Ikediobi, O.N., et al.: Mutation analysis of 24 known cancer genes in the NCI-60 cell line set. Mol Cancer Ther 5(11), 2606–2612 (2006)
8. Martin, Y.C., Kofron, J.L., Traphagen, L.M.: Do structurally similar molecules have similar biological activity? J Med Chem 45(19), 4350–4358 (2002)
9. Dunkel, M., et al.: SuperNatural: a searchable database of available natural compounds. Nucleic Acids Res 34(Database issue), D678–D683 (2006).

# Identification of the Putative Class 3 *R* Genes in *Coffea arabica* from CafEST Database*

Magnólia A. Campos[1,*], Flávia B. Silva[1], Marilia S. Silva[2],
Érika E.V.S. Albuquerque[3], Alexandre M. do Amaral[3], Cristiane C. Teixeira[3],
Angela Mehta[3], and Maria Fátima G. Sá[3]

[1] Universidade Federal de Lavras, Departamento de Biologia, C.P 3037 C.E.P. 37.200-000
Lavras-MG, Brazil
[2] Embrapa Cerrados, Brasília-DF, Brazil
[3] Embrapa Recursos Genéticos e Biotecnologia, Brasília-DF, Brazil
camposma@ufla.br

**Abstract.** Coffee is one of the most important commodities worldwide. For this reason, the sequencing in large scale of expressed sequence tags (ESTs) from different tissues of the coffee tree was performed and resulted in the formation of the Brazilian Coffee Genome EST database (CafEST). There is a raising interest of genetic breeding programs in developing varieties of *Coffea arabica* with increased resistance to nematodes, pests, and diseases. A high number of plant resistance genes (*R* genes) have already been isolated and classified into six categories denoted as class 1 to class 6. In this study, we show results of a screening of the coffee transcriptome database for class 3 LLR/NBS/TIR-like *R* gene related sequences within the *C. arabica* ESTs from the CafEST database. Based on searches for sequence similarities, we selected a total of 293 ESTs coding for class 3 R proteins, putatively related to disease resistance in *C. arabica*. Among these reads, 101 ESTs, representing the RPP4 subclass, were grouped into 56 clusters. We found 93 reads representing the RPP5 subclass, which were grouped into 46 clusters. In addition, we also found 99 reads representing the RPS4 subclass, which were grouped into 54 clusters. However, no matches were found with other subclasses of *R* genes (L, M, N, P, and RPP1) so far. These studies should contribute to the elucidation of the recognition and resistance cascades elicited by *R* genes. These results may provide relevant information to be applied on coffee breeding programs and on the development of new strategies to obtain genetic durable resistance for plants against pathogens, resulting in positive impacts on the coffee agribusiness.

**Keywords:** coffee tree, plant resistance, NBS, LRR, TIR.

## 1 Introduction

Coffee is one of the most important commodities worldwide. Thus, the sequencing in large scale of expressed sequence tags (ESTs) from tissues of coffee trees was performed as an initiative of the Brazilian Consortium for Resources and

---

* Corresponding author.

M.-F. Sagot and M.E.M.T. Walter (Eds.): BSB 2007, LNBI 4643, pp. 171–175, 2007.
© Springer-Verlag Berlin Heidelberg 2007

Development of Coffee together with the AEG-FAPESP and EMBRAPA Genetic Resources and Biotechnology network. Nevertheless, the Brazilian coffee crop is still vulnerable to pest attack, leading to considerable yield losses.

During evolution, plants developed a number of defense mechanisms against pathogens. Besides various preformed barriers, plants also activate resistance at species level (*i.e.*, non-host resistance), race-specific resistance, non-race-specific resistance, and basal resistance mechanisms. Recognition of the presence of pathogens by plants occurs during physical contact between plant and pathogen either on the plant external contact surface or inside the plant tissues (Vidhyasekaran, 1997). Therefore, the recognition of the pathogen by the plant, which is mediated by protein resistance receptors (R proteins), is essential for the induction of local defense responses. A high number of plant resistance genes (*R* genes) has already been isolated and was classified into six categories denoted as class 1 to class 6. All the class 3 *R* genes, which present TIR-NBS-LRR domains, possess an N-terminal domain resembling the cytoplasmic signaling domain of the Toll and Interleukin-1 (TIR) transmembrane receptors, such as L, RPS4, RPP1, RPP5, and N (Whitham *et al.*, 1994; Meyers *et al.*, 1999; Dangl and Jones, 2001; Tör *et al.*, 2004). TIR motifs are ancient. Drosophila Toll and mammalian Toll-like receptors (TLRs) recognize PAMPs (pathogen-associated molecular pattern) through the extracellular LLR domain and transduce the PAMP signal through TIR domain. Similarly, induced non-host resistance in plants is comparable to animal innate immunity, which activates pathogen resistance following host recognition of these PAMPs general elicitors (Jones and Takemoto, 2004).

We are performing a data-mining-based identification of plant disease *R* genes in the Brazilian Coffee Genome EST database (CafEST). Among over 2,400 reads have already been found within this database. In this work, we report the existence of a large number of class 3 *R* genes within the *Coffea arabica* genome.


## 2  Materials and Methods

To search homologous genes encoding class 3 LLR/NBS/TIR-like R proteins in *C. arabica* genome, comparison analyses were performed within the CafEST database (http://www.lge.ibi.unicamp.br/cafe) using the Basic Local Alignment Tool (BLASTn) program (Altschul *et al.*, 1997) and nucleotide sequences of each of the well described subclasses (L, M, N, P, RPS4, RPP1, RPP4, and RPP5) found in public databases (www.nbci.nlm.nih.gov) as query. The use of the cut-off value of $1e\text{-}10^{-4}$ and BLOSUM62 matrix criteria led to the selection of 293 EST reads from CafEST, which comprises 153,739 valid reads. EST clusters were built for each subclass separately, through the alignment by using the Contig Assembly Program (CAP3) (Huang and Madan, 1999). Consensus sequences of each cluster were compared with the amino acid sequences from protein homologous sequences from the Genbank, by using the BLASTx program (Altschul *et al.*, 1997).

All *C. arabica* sequences used in this work, corresponding to sequenced EST reads and cluster consensus, were obtained from the Brazilian Coffee Genome EST database (CafEST), which was built from cDNA libraries for different genotypes, organs (leaf, stem, fruit, flower, and root) or growth and stress conditions, as described in detail elsewhere (Vieira et al., 2006).

## 3 Results and Discussion

In this work, we report the presence of *R* genes encoding class 3 LLR/NBS/TIR-like R proteins within the *C. arabica* ESTs from the CafEST database, which are involved in signal perception of non-self effector molecules. Based on sequence similarities in homologue searches, using the BLASTn program, we selected a total of 293 ESTs coding for putative class 3 R proteins related to disease resistance in *C. arabica* (Table 1). Among these reads, 101 ESTs representing the RPP4 subclass were grouped into 56 clusters. We found 93 reads representing the RPP5 subclass, which were grouped into 46 clusters. *Arabidopsis RPP4* is a member of the *RPP5* multigene family of TIR-NB-LRR genes and related to downy mildew resistance through multiple signaling components (van der Biezenet et al., 2002). In addition, we also found 99 reads of RPS4 subclass, which were grouped into 54 clusters. The *Arabidopsis RPS4* bacterial-resistance gene is a member of the TIR-NBS-LRR family of disease resistance genes whose protein recognizes the *avrRps4* avirulence gene product from *Pseudomonas syringae* pv. *pisi* (Hinsch and Staskawicz, 1996; Gassmann *et al.*, 1999). However, no hit was found when nucleotide sequences of other subclasses of *R* genes (L, M, N, P, and RPP1) were used as probes for CafEST database mining.

**Table 1.** Number of putative class 3 *R* genes found within CafEST database in the *Coffea arabica* genome

| Gene name | Number of reads | Number of clusters | |
|-----------|-----------------|--------------------|--------------------|
| | | Contigs | Singlets |
| RPP4 | 101 | 19 | 37 |
| RPP5 | 93 | 17 | 29 |
| RPS4 | 99 | 18 | 36 |

Interestingly, within the clusters of RPP4 (Table 2) we found that the contigs C1, C3, C5, C13, C16 and C18 are exclusively formed by reads from a single coffee library (data not shown): C1 is formed by 2 reads expressed in leaves treated with araquidonic acid; C3 is formed by 2 reads from leaves not treated with Bion; C5 is composed by 5 reads from *Xylella fastidiosa*-infected branches, what suggests the involvement of this contig sequence in the basal coffee resistance against this pathogenic bacteria; C13 is composed by 3 reads originated from coffee cells maintained in saline medium; C16 is formed by 2 reads from root tissue and suspension of cells in the presence of Al and the C17 is composed by 3 reads originated from seeds during germination.

**Table 2.** Distribution of putative *Coffea arabica RPS4*-ESTs in clusters and homology analysis

| Contig | Read length (nt) | Number of reads | Best Blast hit | | | |
|---|---|---|---|---|---|---|
| | | | Organism | Access number | E-value | Similarity |
| C1 | 831 | 2 | *Solanum tuberosum* | gb\|AAP44392.1 | 2E-56 | 131/153 (84%) |
| C2 | 1579 | 9 | *Oryza sativa* | ref\|NP_915900.1 | 2E-16 | 156/378 (41%) |
| C3 | 736 | 2 | *Arabidopsis thaliana* | ref\|NP_200956.1 | 5E-68 | 161/245 (65%) |
| C4 | 1556 | 7 | *Oryza sativa* | ref\|NP_001067154.1 | 1E-141 | 312/408 (76%) |
| C5 | 779 | 3 | *Populus trichocarpa* | gb\|ABF81421.1 | 1E-20 | 119/249 (47%) |
| C6 | 769 | 3 | *Cucumis melo* | gb\|AAT77096.1 | 2E-15 | 88/182 (48%) |
| C7 | 1282 | 3 | *Populus trichocarpa* | gb\|ABF81421.1 | 1E-27 | 158/334 (46%) |
| C8 | 734 | 3 | *Cucumis melo* | gb\|AAT77098.1 | 6E-08 | 74/176 (42%) |
| C9 | 999 | 2 | *Arabidopsis thaliana* | gb\|ABG00804.1 | 1E-58 | 171/281 (59%) |
| C10 | 900 | 2 | *Medicago truncatula* | gb\|ABE84400.1 | 3E-29 | 95/164 (57%) |
| C11 | 918 | 7 | *Capsicum annuum* | gb\|AAN62015.2 | 2E-68 | 161/197 (81%) |
| C12 | 1552 | 4 | *Glycine max* | gb\|AAR19098.1 | 4E-29 | 175/397 (44%) |
| C13 | 862 | 3 | *Capsicum annuum* | gb\|AAN62015.2 | 5E-52 | 151/179 (84%) |
| C14 | 922 | 2 | *Arabidopsis thaliana* | ref\|NP_176532.2 | 9E-76 | 182/257 (70%) |
| C15 | 1003 | 2 | *Arabidopsis thaliana* | gb\|AAM13028.1 | 7E-67 | 167/256 (64%) |
| C16 | 685 | 2 | *Solanum tarijense* | gb\|AAR29076.1 | 6E-28 | 122/215 (56%) |
| C17 | 798 | 3 | *Solanum tuberosum* | gb\|AAW48301.1 | 6E-09 | 59/113 (52%) |
| C18 | 784 | 3 | *Nicotiana benthamiana* | gb\|AAY54606.1 | 6E-53 | 145/214 (67%) |

## 4  Concluding Remarks

In this study, the criteria used to screen the CafEST database for ESTs from *C. arabica* that code for *R* genes class 3 allowed the identification of various putative homologous genes to *RPP4*, *RPP5*, and *RPS4*. Taking into consideration that coffee plants are perennial, the data shown provide relevant information to be used for classical breeding programs as well as for the development of new approaches to reach coffee durable resistance against pathogens, leading to positive impact on the coffee agribusiness.

## Acknowledgments

The authors acknowledge support received from Embrapa Recursos Genéticos e Biotecnologia, Embrapa Café and Consórcio Brasileiro de Pesquisa e Desenvolvimento do Café. M.A.C. has financial support given by CAPES/ PRODOC/ UFLA, Brazil.

## References

Altschul, S., Madden, T., Schaffer, A., Zhang, J., Zhang, Z., Mille, W., Lipman, D.J.: Gapped BLAST and PSI-BLAST: A new generation of protein database search programs. Nucleic Acids Res. 25, 3389–3402 (1997)

Dangl, J.L., Jones, J.D.: Plant pathogens and integrated defense responses to infection. Nature 411, 826–833 (2001)

Gassmann, W., Hinsch, M.E., Staskawicz, B.J.: The Arabidopsis RPS4 bacterial-resistance gene is a member of the TIR-NBS-LRR family of disease resistance genes. Plant J. 20, 265–277 (1999)

Hinsch, M., Staskawicz, B.: Identification of a new Arabidopsis disease resistance locus, RPS4, and cloning of the corresponding avirulence gene, avrRps4, from Pseudomonas syringae pv. pisi. Mol Plant Microbe Interact 9, 55–6177 (1996)

Huang, X., Madan, A.: CAP3: A DNA Sequence Assembly Program. Genome Res 9, 868–877 (1999)

Jones, D.A., Takemoto, D.: Plant innate immunity - direct and indirect recognition of general and specific pathogen-associated molecules. Curr Opin Immun. 16, 48–62 (2004)

Meyers, B.C., Diekcman, A.W., Michelmore, R.W., Sivaramakrishnan, S., Sobral, B.W., Young, N.D.: Plant disease resistance genes encode members of an ancient and diverse protein family within the nucleotide-biding superfamily. Plant J 20, 317–332 (1999)

Tör, M., Brown, D., Cooper, A., Woods-Tör, A., Sjolander, K., Jones, J.D.G., Holub, E.: Arabidopsis downy mildew resistance gene RPP27 encodes a receptor-like protein similar to CLAVAT2 and tomato Cf-9. Plant Physiol. 135, 1–13 (2004)

van der Biezen, E.A., Freddie, C.T., Kahn, K., Parker, J.E., Jones, J.D.: Arabidopsis RPP4 is a member of the RPP5 multigene family of TIR-NB-LRR genes and confers downy mildew resistance through multiple signaling components. Plant J. 29, 439–451 (2002)

Vieira, L.A.A., Colombo, C., Moraes, A., Mehta, A., et al.: Brazilian coffee genome project: an EST-based genomic resource. Braz J Plant Physiol. 18(1), 95–108 (2006)

Videhyasekaran, P.: Fungal pathogenesis in plants and crops. Molecular biology and host defense mechanisms, p. 553. Dekker Inc, New York (1997)

Whitham, S., Dinesh-Kumar, S.P., Choi, D., Hehl, R., Corr, C., Baker, B.: The Product of the tobacco mosaic virus resistance gene N: similarity to toll and thr interleukin-1 receptor. Cell 78, 1101–1115 (1994)

# Author Index

# Lecture Notes in Bioinformatics

Vol. 3240: I. Jonassen, J. Kim (Eds.), Algorithms in Bioinformatics. IX, 476 pages. 2004.

Vol. 3082: V. Danos, V. Schachter (Eds.), Computational Methods in Systems Biology. IX, 280 pages. 2005.

Vol. 2994: E. Rahm (Ed.), Data Integration in the Life Sciences. X, 221 pages. 2004.

Vol. 2983: S. Istrail, M.S. Waterman, A. Clark (Eds.), Computational Methods for SNPs and Haplotype Inference. IX, 153 pages. 2004.

Vol. 2812: G. Benson, R.D.M. Page (Eds.), Algorithms in Bioinformatics. X, 528 pages. 2003.

Vol. 2666: C. Guerra, S. Istrail (Eds.), Mathematical Methods for Protein Structure Analysis and Design. XI, 157 pages. 2003.